Construction Reports 1944–98

Construction Reports 1944–98

Edited by

Mike Murray and David Langford
Department of Architecture
and Building Science,
University of Strathclyde

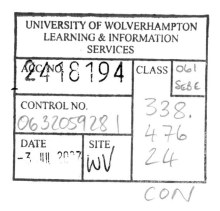
Blackwell
Science

© 2003 by Blackwell Science Ltd,
a Blackwell Publishing Company
Editorial Offices:
Osney Mead, Oxford OX2 0EL, UK
   Tel: +44 (0)1865 206206
Blackwell Science, Inc., 350 Main Street,
Malden, MA 02148-5018, USA
   Tel: +1 781 388 8250
Iowa State Press, a Blackwell Publishing Company,
2121 State Avenue, Ames, Iowa
50014-8300, USA
   Tel: +1 515 292 0140
Blackwell Publishing Asia Pty,
550 Swanston Street, Carlton South,
Victoria 3053, Australia
   Tel: +61 (0)3 9347 0300
Blackwell Wissenschafts Verlag,
Kurfüstendamm 57, 10707 Berlin, Germany
   Tel: +49 (0)30 32 79 060

First published 2003 by Blackwell Science Ltd

Library of Congress
Cataloging-in-Publication Data is available

0632–05928–1

A catalogue record for this title is available from the
British Library

Set in 10/12pt Palatino by
Bookcraft Ltd, Stroud, Gloucestershire
Printed and bound in Great Britain by
MPG Books Ltd, Bodmin, Cornwall

For further information on
Blackwell Science, visit our website:
www.blackwell-science.com

Contents

Contributors

Dr David Boyd is Deputy Head of the School of Property and Construction at the University of Central England in Birmingham. Although his background is in science and technology, Dr Boyd is better known for his sociological and management perceptions of the industry. He has undertaken extensive action research into management in the construction industry. From this he has investigated the ways in which complexity and uncertainty are perceived and managed so that new models of the operation of projects can be created. He believes that any change or innovation involves both organisations and individuals in learning, and this has been his approach in analysing, promoting and developing new thinking in the industry. He has devised and run a number of innovative Masters courses using ideas of learning to learn and reflection in practice.

Dr Daniel Cahill is a Lecturer in the School of the Built Environment at Heriot-Watt University, Edinburgh, specialising in business management in the construction industry. He is also the Course Director of the MSc in Construction Project Management at Heriot-Watt University. He has a good industrial background, having spent several years working in professional practice and contracting prior to joining Heriot-Watt University. His research interests are design information management and strategic management. Design information management is concerned with the use of information in construction problem-solving, using both traditional methods and new techniques. His PhD research was on the utilisation of information in architectural design drawings. Strategic management research focuses the implications of changes in client demand for the construction service. This is primarily based around workshops and industrial liaison, with particular emphasis on the impact of technological change on the need for space/buildings and how this affects the overall construction market, the implications of Latham, Egan, PFI/PPP and partnering and the resulting organisational and cultural changes in the industry.

Dr John Connaughton is a Partner in Davis Langdon Consultancy (DLC), the specialist research and management consultancy division of Davis Langdon and Everest. John has over 20 years experience in the construction industry, 15 of which have been spent in the management and direction of research and consultancy projects. He has undertaken research and consultancy for a wide range of clients including the UK Department of Trade and Industry, HM Treasury, the Ministry of Defence, the RICS, RIBA, ACE and CIRIA, among others. Currently, John is responsible for much of DLC's research and consultancy work on the strategic management of construction, and on improving the building procurement process. He writes and lectures widely in the areas of construction management and procurement, client demand, the organisation of professional knowledge and responsibility and international construction. He is the author of a range of reports and guidance for both clients and consultants, including *Benefit Trading: A Practical Guide for Construction* (with Malcolm Potter) and *Value Management in Construction: A Clients' Guide* (with Stuart Green). John is a member of the Movement for Innovation (M4I) Board, and is an independent board member of Genesis Housing Group. He was formerly Chairman of the Construction Industry Council's Innovation and Research Committee and the Environment Committee.

Dr Stuart Green is a Senior Lecturer in the School of Construction Management and Engineering at The University of Reading. He is Course Director of the Reading MSc in Project Management. Previously he worked in contracting for several years before gaining design experience with an engineering consultancy. He is a Chartered Civil Engineer and Chartered Builder. He has published extensively on a wide range of topics and is especially well known for his work on value management. Current funded research projects include *Knowledge Sharing between Aerospace and Construction* and the *Human Resource Management Implications of Lean Construction*. Much of his work is critical in orientation. He is well known throughout the UK construction industry as a dissenter to the Egan agenda. Numerous publications have challenged currently accepted notions of 'best practice'. Recent targets have included partnering, lean construction and process improvement.

Professor Cliff Hardcastle holds an MSc by Research and a PhD. He is a Member of the Chartered Institute of Building and the Association of Cost Engineers. He is currently Dean of the School of the Built and Natural Environment at Glasgow Caledonian University. Cliff's previous research work in construction procurement includes the potential of integrated databases for information exchange in the construction industry, and the analysis of approaches to procurement and cost control of petrochemical works. He is currently involved in a number of research projects including *Building Procurement Decision-making* and *Decision-making and Risk in PFI Projects*. He also holds a major grant together with Strathclyde University and the

Building Research Establishment (Scotland) for the establishment of the Centre for Advanced Built Environment Research. He is a member of the RICS Research Advisory Board, an Advisor to the Malaysian Board of Quantity Surveyors and referee for a number of international research councils including the United Kingdom, South Africa and Hong Kong.

Dr Patricia Hillebrandt has for many years been a Senior Visiting Research Fellow at the University of Reading and is now engaged on a research project on the cost of procurement in construction. She has since 1967 been an economic and management consultant on the construction industry. Previously she was employed as economist to the contracting company Richard Costain Ltd, at the National Economic Development Office and as Senior Lecturer at the Bartlett School of Architecture and Planning, University College London. She has been a consultant to many international and Government organizations, including the UK Ministry of Overseas Development, the World Bank, The International Labour Organization, the European Union and DANIDA, the Danish aid organisation. She has therefore worked extensively abroad, notably in Egypt, Sri Lanka, Russia and Latvia. She is the author of several publications on the construction industry, including *Analysis of the British Construction Industry* and *Economic Theory and the Construction Industry*. She is also the author or editor (with others) of several books on the way contractors take their management decisions.

Dr Graham Ive is Senior Lecturer in the Economics of Construction at the Bartlett School of Graduate Studies, University College London (UCL), where he has been for many years. He is the author of two recent books, *The Economics of the Modern Construction Sector* and *The Economics of the Modern Construction Firm*. He is also coauthor of two major studies of PFI, both the outcome of research projects undertaken with the Construction Industry Council for DETR under the Partners in Innovation programme [*Constructors' Key Guide to PFI* (1998) and *The Role of Cost Saving and Innovation in PFI Projects* (2000)]. He has been an economic adviser to the CIC, and is a member of its Economic Task Force. In addition to his own research on PFI, he is currently supervising three PhDs in this topic. At UCL he directs the MSc programme in Construction Economics and Management, which has between 25 and 30 post-experience graduate students each year, drawn from all the built environment professionals and from all around the world. He learnt his economics at Cambridge and is a member of the Post-Keynesian Economics Group and the Association of Heterodox Economists.

Peter Kennedy is a Chartered Builder who started his working life in site engineering with a contractor. He then progressed through various roles as planning engineer, site manager and quantity surveyor. After 11 years in contracting, he took an MSc in Construction Management and moved into the academic world. He is currently a Senior Lecturer in Construction

Management in the School of the Built and Natural Environment at Glasgow Caledonian University. Peter's research interests include the management of conflict in construction and dispute resolution. Following the introduction of statutory adjudication in 1998, he established the Adjudication Reporting Centre which collects data from the Adjudicator Nominating Bodies and adjudicators themselves to provide information on the causes of conflict, the parties in dispute, the sums of money in dispute and trends in the growth of adjudication. In recent years he has advised the Government of New Zealand on the content of their proposed legislation in the field of security of payments and adjudication of disputes.

Professor David Langford is the Director of the Graduate School in the Department of Architecture and Building Science at Strathclyde University, having had experience of working in the construction industry before taking up an academic appointment. During his academic career he has published, in conjunction with colleagues, several books and has contributed to many academic journals and conferences. He has taught and consulted throughout the world.

Professor Steven Male holds the Balfour Beatty Chair in Building Engineering and Construction Management, School of Civil Engineering, University of Leeds. His research and teaching interests include strategic management in construction, supply chain management, value management and value engineering. He has led research projects under the EPSRC IMI programme *Construction as a Manufacturing Process*, with the DETR and DTI and within the European Union 4th and 5th Frameworks. He is a visiting Professor in the Department of Civil Engineering, University of Chile. He works closely with industry and has undertaken a range of research, training and consultancy studies with construction corporations, construction consultancy firms and blue chip and Government clients.

Dr Lawrence Mbugua is a Consultant with Davis Langdon Consultancy (DLC). At DLC, Lawrence is currently managing a portfolio of research projects funded by the Department of Trade and Industry aimed at improving business performance and encouraging best practice within UK construction companies. He has a PhD in Construction Business Performance Measurement from the University of Wolverhampton, where he is a Visiting Lecturer. Prior to that, Lawrence had undertaken an MSc in Property Investment and Development at UMIST and a BA in Building Economics and Management at the University of Nairobi. Lawrence has also worked as an investment analyst at the Nairobi Stock Exchange and was involved in the set-up of the Kenya Agricultural Commodity Exchange. His professional interests include performance measurement, benchmarking, strategy, construction economics and property investment and development.

Krisen Moodley is Programme Leader for MSc courses in Construction Management in the School of Civil Engineering, University of Leeds. After graduating from Natal University, his initial employment was as a quantity surveyor with Farrow Laing in Southern Africa, before his first academic appointment at Heriot-Watt University. He spent 4 years at Heriot-Watt before joining Leeds in 1994. His research interests are concerned with the strategic business relationships between organisations and their projects. Other specialist research interests include: procurement, project management and corporate responsibility. He has contributed to many publications and is coauthor of the book *Corporate Communications in Construction*.

Mike Murray is a Lecturer in Construction Management within the Department of Architecture and Building Science at the University of Strathclyde. He is currently at the closing stage of his PhD research and holds a first-class honours degree and MSc in Construction Management. He has lectured at three Scottish universities (The Robert Gordon University, Heriot-Watt and currently at Strathclyde) and has developed a pragmatic approach to both research and lecturing. As well as delivering research papers at both UK and overseas conferences, Mike has been involved in presentations during workshops with contractors Costain, Wates and Gleeson. His career within the building industry began with an apprenticeship in 1980 and his future research aspirations include a study of the 'stigmatised' perceptions of occupations within the building trades.

Dr Christopher Preece trained as a contracts planner and is now Course Leader on the MEng/BEng Civil Engineering with Construction Management programme in the School of Civil Engineering, University of Leeds. As a member of the Construction Management Group, he is directing research in the field of business, marketing and quality management in the construction industry. The focus of these studies is on increasing client satisfaction and improving the business culture of the construction industry through more effective and competitive business processes. He is research coordinator for CIB W065 TG1 International Construction Marketing. Chris has published and presented more than 30 papers both nationally and internationally. His recent books include *Corporate Communications in Construction* and a contribution to *Strategic Management in Construction*.

Dr Marie-Cécile Puybaraud is a Lecturer in the School of the Built Environment at Heriot-Watt University, Edinburgh. She is also the Course Director of the MSc in Facilities Management and leads the Facilities Management Research Group at Heriot-Watt University. Her expertise is in fire safety management and business continuity planning in facilities. She also specialises in aspects of the legislation for the construction industry in the EU, contractual law and administration of UK and international construction contracts and strategic procurement. Marie-Cécile's research interests focus on the role of management in fire safety scenarios and fire safety management in facilities. She obtained her PhD in July 2001 and was awarded the MacFarlane Medal for best doctorate of the year, excellence in research and major contribution to research. Her groundbreaking and unique research in the field of fire safety on construction sites led to the creation of a *Fire Safety Management Model* for the construction industry through an interactive CD-ROM. She is currently extending her expertise by collaborating with fire safety experts in the UK, France, Australia and China. Marie-Cécile is an associate member of the Chartered Institute of Building and the British Institute of Facilities Management and a member of the Institute of Learning and Teaching.

Dr John Tookey is a Lecturer in Construction Logistics and Marketing at Glasgow Caledonian University. He graduated in 1993 with a BSc in Industrial Technology and subsequently obtained his PhD in Industrial Engineering from the University of Bradford in 1998. His main research interests are in supply chain management (SCM) and logistics within a construction context. This has grown out of his PhD research (looking at SCM in the aerospace industry) and more recently his post-doctoral research in construction procurement at Glasgow Caledonian University.

Alan Wild is Senior Lecturer in the Division of Management Studies in the University of Central England Business School. For the past 12 years he has worked on a Masters programme for built environment practitioners around related concepts of the emerging theory of construction and property and Educating for uncertainty. He publishes actively through ARCOM and has had an article accepted by *Engineering, Construction and Architectural Management*. Presently he is researching reports and publications on construction from the 1950s and 1960s. He has tutored for The Open University and Henley Management College on their Distance Learning programmes, and for the Warwick Manufacturing Group. Alan was a member of the Project Support Group for the Tavistock Institute's research on effective learning networks in construction, conducted on behalf of CIRIA from 1996 to 1997. His wider interests lie in understanding the impossibility of construction, uncertainty in management, the development of managers as reflective practitioners and project management more generally.

Foreword

Clive Thomas Cain CBE ARIBA

This book is essential reading for all those concerned for the future well-being of the construction industry, from the construction professionals within client organisations, through academics, consultants, construction contractors, and specialist suppliers to manufacturers. It proves beyond doubt that a continuous stream of reports over the last 50 years has raised the same concerns over the industry's poor performance and has given the same warning of the cost consequences this has on the competitiveness of other sectors. It also proves the industry's continuing unwillingness to accept the message of the reports and radically change its structure and its culture in order to improve its performance and to deliver better value to its end-user clients.

The conclusion of the book ought to be mounted in front of every key figure in the industry, as a constant reminder of what really needs to be improved if the industry is serious about addressing the business needs of its end-user clients.

Clive Cain was the Quality Director at Defence Estates and was also responsible for the pioneering Building Down Barriers development project. Prior to his early retirement, he was a member of the Movement for Innovation Board and was the Defence Estates' representative on the Design Build Foundation and the Construction Round Table. Clive currently coaches in supply chain management (lean construction) and has written a definitive booklet *A Guide to Best Practice in Construction Procurement* published by the UK Construction Best Practice Programme.

Preface

This book is aimed at graduates, researchers and practitioners in all fields of the built environment and serves as a record of the manner in which the UK Government has intervened in the post-World War II construction industry. The 12 reports reviewed in this book begin with the Simon Committee report of 1944 and conclude with the 1998 *Rethinking Construction* (Egan) report. Most, though not all Government reports that have tackled the state of the industry are evaluated.

One of these, the Tavistock report, was not commissioned by the Government but was seen as sufficiently influential to warrant inclusion in our list. Others have not made the review. These include the Bronowski report (1950) which considered standardisation and the opportunities for prefabrication, a subject still resonant in the early 2000s. The Phelps-Brown report (1968) was an important enquiry into matters concerning labour-only subcontracting. This was an important report but has been omitted because its scope was limited to one aspect of the industry – labour only subcontractors – and our concerns were related to the larger picture of the performance of the industry as a whole. Other omissions include the O'Reilly report (1987) relating to briefing. The Foster report (1988) and Atkins report (1993) have also been excluded since they were not specifically Government reports. Our excuse for not incorporating these texts is the limited space available and the need to exercise some selectivity.

The editors invited a selection of academics to contribute to this text, and therefore the book offers an eclectic mix of reviews. In some cases, particularly with the earlier reports, it has been difficult to ascertain their impact on the industry. What is known, however, is that for several decades after World War II the industry continued to adopt a fragmented construction process that involved sequential procurement, in spite of the recommendations given in report after report. The reader will learn that, notwithstanding a gap of 54 years between the first and last reports reviewed, the message is strikingly similar: construction project teams must work together in true partnership and not as groups of disparate professions. Thus, the call for early involvement of subcontractors can be seen in the earlier reports, well before the 1990s management fad of supply chain management.

Chapter 1

Introduction

David Langford & Mike Murray

1.1   A review of reviews

The need for a textbook that reviews the manner in which the UK Government has intervened in the construction industry is made clear in one recent report. The aptly named *Modernising Construction* report[1] confirms what all involved in construction have known for far too long, that is, the overwhelming failure to act on recommendations made in successive construction reports since 1944:

> 'There are still too many clients, consultants and constructors who see partnering as an alien or threatening process. They could usefully reflect on how badly they have been served by traditional methods. If all had been well with the construction industry, there would have been no need for the long stream of reports on its performance since Simon in 1944.'

The coverage of reports in this text does indeed start with Simon in 1944[2], but closes with a review of the 1998 *Rethinking Construction* report[3]. The writing and editing of any book requires a time line to be drawn through the process of bringing the book to publication. However, let us look ahead before the start of our half-century review. At the time of writing, the follow-up to Egan's 1998 report is at draft stage. Expected in September 2002, it has been entitled *Accelerating Change* and reflects the desire to speed up the cultural changes recommended in the 1998 report. This clear desire to maintain the momentum of change is most evidently lacking in the pre-1994 reports reviewed in this book. The willingness of industry – clients, consultants and constructors alike – to embrace change has no definitive explanation. Indeed, given the historical nature of the reports reviewed here, a significant amount of anecdotal evidence and speculation is offered by each author as to why this may be so. Currently, the plethora of Eganised information cascading throughout industry shows no sign of stopping. There is a sad lack of such literature on earlier reports. This, then, is largely the purpose of this text, to

1

provide a critique of earlier reports, brought together in one document and acting as a ready-reference for those readers who wish to examine the historical development of the post-World War II construction industry.

Since World War II there have been many Government interventions in the construction industry. The relationship between Government and industry has never been easy; for much of the period under review the construction industry has relied on Government for a steady stream of work, and the industry has resented Government power to switch on or switch off the workload as economic conditions have varied. Yet the influence is subtler than this crude economic leverage would suggest. Governments have sought to influence behaviour, procurement, performance, relationships between clients, designers, contractors and subcontractors, contractual arrangements, briefing procedures, etc. The list is long.

This introduction is broken down into three parts: the first deals with the Government ministries that have had an interest in construction, the second identifies the drivers of the report and the third evaluates their impact. These themes will be examined in greater detail within each chapter and revisited in the conclusions.

1.2   The ministries and the Ministers

In the period between World War II and January 2002 there have been 13 Governments. These have been led by 11 Prime Ministers and have included 37 Secretaries of State and Ministers of State who have had an interest in the construction industry. The task of tracing the political responsibility for construction is difficult. As the tome *British Political Facts 1900–2000*[4] points out, the Ministers responsible for environmental matters have changed constantly. The traditional political home for construction in the post-World War II period was the Ministry of Building and Public Works, but the concept of having a centrally directed effort in Britain fell away in the 1960s, and the political home for construction then veered between Ministers of Construction, Housing, Local Government, Water, Inner Cities and Environment and is now the Ministry of Trade and Industry. Naturally, 'Transport' also had an overview of construction projects that involved roads and, as Butler & Butler point out[4], ministries overseeing construction have 'variously combined, often for short periods and without any change in the ministerial team'. It was not until the Heath Government of 1970–74 that the first Minister of Housing and Construction (Julian Amery) was appointed.

From an inspection of the political careers of the Ministers with an interest in construction (Table 1.1), one can see that it is an important if impermanent position. What is also noticeable is that few of the Ministers have had any connection with construction. Nicholas Ridley, a Conservative Environment Minister, was educated as a civil engineer, John Marples, a Conservative Minister of Transport, was the founder of the construction giant Marples

Table 1.1 Ministers with a construction remit.

Report	Year	Minister	Political affiliation	Department
Simon	1944	Lord Woolton	National Government	Reconstruction
Phillips, received by Richard Stokes	1950	Charles Key	Labour	Ministry of Works
Emmerson	1962	Henry Brook	Conservative	Housing and Local Government
Banwell	1964	Henry Brook	Conservative	Housing and Local Government
Received by		Richard Crossman	Labour	Housing and Local Government
Tavistock	1965/ 1966	Not commissioned by Government		
What's Wrong on Site?	1970	Richard Crossman	Labour	Housing and Local Government
Received by		Julian Amery	Conservative	Housing and Construction
Wood	1975	Tony Crosland	Labour	Environment
Faster Building for Industry	1983	Michael Heseltine	Conservative	Environment
Received by		Tom King	Conservative	Environment
Faster Building for Commerce	1988	Nick Ridley	Conservative	Environment
Latham	1994	John Gummer	Conservative	Environment
Technology Foresight	1995		Conservative	Not construction specific but driven by the Office of Science and Technology based in the Cabinet Office
Egan	1998	John Prescott	Labour	Environment, Transport and Regions

Ridgeway and Peter Walker, again a Conservative Minister for the Environment, had strong interests in property and construction, and the Slater Walker construction firm bore his name. On the Labour side, only John Gilbert has held a boardroom position with a construction firm.

1.3   The drivers of the reports

On inspecting the reports that are reviewed in this book, it would appear that the intervention for most of the Government reports (Tavistock is an exception) has been driven by only two groups of powerful clients wanting a better performing industry to serve their needs: in the period 1944–80 these would have been powerful Government or parastatal clients, while the 1980–2000 period became the era dominated by powerful private clients or construction employers wishing to redraft the boundaries and rules for conducting business between themselves and the state. Testing each report against the client or constructor proposition, we can see a picture emerging of the driving forces behind the reports (Table 1.2).

1.4   Recurring themes

In reviewing the big picture of the reports, all of them have in some way or other encouraged a set of changing relationships between the parties to the construction process. The main objective of such changed relationships has been the improvement of performance for the industry. What is left unsaid in every report is precisely who benefits from this performance improvement. To argue against the benefits of performance improvements is naive, especially if such improvements lend themselves to lower costs, greater functionality, shorter construction times, etc. All of these issues are recurring themes. Yet it is assumed that the benefits of any changes in the construction process will accrue to the principle elite members of the construction team – the designers, construction firms or clients. When the client is the public sector, clearly there is public benefit, but public sector expenditure in construction, while still substantial, has fallen from around 50% in the mid-1970s to just over 20%, and so the public gain has diminished over the years. In spite of this drop, Government departments and their agencies accounted for £7.5 billion (from a total UK spend of £65 billion) of work in 2000. Indeed, a spending review carried out in 2000 proposed additional investment of up to £19 billion for improvement in transport, schools and hospitals[1].

The theme of procurement (Simon, Emmerson, Banwell, Wood, Latham and Egan) provides a constant thread through the post-World War II years. Each of these reports describes a condition of continuity of work for the industry as being a desirable outcome. It is perhaps ironic that the continuity of work for the few selected Private Finance Initiative (PFI) contractors (those with balance sheets capable of supporting a series of PFI bids) now comes from their operation and maintenance of the facilities they have built. Indeed, the continuity of income over a concession period of 25–30 years is preferable to contractors who previously may have depended on sporadic public spending over the past 40 years. Moreover, to some extent the language in

Table 1.2 Drivers identified in reports.

Report	Theme	Driver
Simon	Placing of public contracts	Constructors want less bureaucratic tendering in Government contracts. Escape from competitive tendering
Phillips	Organisation and efficiency of the building industry	Public clients seek better performance from the industry through improvements in labour productivity and the management of the construction process
Emmerson	Greater integration of the design and construction process	Constructors want a continuous stream of work less dependent on open tenders and incomplete design information
Banwell	Management of the building process	Constructors look to Government to regulate the placing of contracts. Public contracts negotiable
What's Wrong on Site?	Industrial relations on large sites	Clients want better control of projects and industrial relations in particular
Wood	Placing of public contracts via package deals	Constructors want more negotiated work and final contracts. Architects alarmed
Faster Buildings for Industry	Productivity in building factories and warehouses	Property developers (clients) want US-type construction times for UK
Faster Building for Commerce	Productivity in commercial construction	Property developers (clients) want faster construction times for office blocks
Latham	Relationships between the parties to the construction process	Both clients and constructors gain: clients through better performance, constructors through better cash management
Technology Foresight	Return to an industry planning model not seen since the 1960s	Political, social and technical alignment of a changed agenda set by Government. Prepares the ground for Egan
Egan	Performance and productivity of the industry	Clients want and get greater authority over the constructors

this post-Egan industry may also have changed, with 'continuity' being replaced by 'sustainability'.

This stability is to be reinforced by a move away from a system where design is separated from building. The traditional system of procurement was derided for not delivering to clients the performance for which the industry is capable. Cain[5] has recently argued that many of the reports on the industry have failed to have an impact on performance because clients have continued to reinforce fragmentation by using a sequential procurement process.

The issue of performance also pervades the reports (Simon, Emmerson, Banwell, Wood, *Faster Building for Industry, Faster Building for Commerce,*

What's Wrong on Site?, Latham, Egan). The earlier reports provide vague generalisations about what performance improvement is possible and how it may be achieved. The more recent reports become more hard edged, with specific targets for time and cost saving by set dates. The later reports, Latham and Egan, specified targets and proposed mechanisms whereby performance measurement could take place. The use of key performance indicators (KPIs) is one such mechanism of cascading a culture of change throughout the industry.

Organisations such as Movement for Innovation (M4I) and the Construction Best Practice Programme were agencies given a role in driving through the changes. The agencies have provided a new environment in which change is promulgated from major reports. Paradoxically, another recurring theme has been 'plus ça change, plus c'est la même chose'. In the 13 reports covered in the 58 years since Simon, a noticeable feature of the reports is the resurrection of old issues. What has changed is the willingness to provide vehicles to implement change. In more recent reports the industry and Government have developed organisational learning skills such that reports are acted upon rather than left on the shelf. The evidence for this change comes from reviewing the end-of-year commentary in the magazine *Building* (Table 1.3). One can gauge the impact of reports by exploring their coverage in the review of the year. This did not start until 1975 with the Wood report (incidentally, in 1964 and 1970 the dateline on the last *Building* issue of the year is 25 December). It can be seen that it is only the later reports that capture the imagination of the industry.

Table 1.3 Journal commentary on reports.

Report	*Building* end-of-year commentary
Wood, 1975	The Wood report is mentioned through a piece entitled 'Who made news in 1975'. *Building* featured a story of each month of the year. The report does not feature in the 12 bylines
Faster Building for Industry, 1983	No mention in the end-of-year leader
	No mention in the review of the year
Faster Building for Commerce, 1988	No mention in the end-of-year leader
	No mention in the month-by-month news review (this carried 40 side-bar panels)
Latham, 1994	Opinion column reflects on the ways and means of achieving Latham targets
	Review of the year features Latham
Technology Foresight, 1995	No mention in the end-of-year leader
	No mention in the review of the year
Egan, 1998	Egan features in the review of the year

Thus, for the last half-decade the industry has been the subject of a Government report every 3–5 years. Can the series of works be said to have shaped the industry? In the earlier reports the content was largely research based, either by empirical enquiry or by offering new sets of insights into the workings of the industry and the relationships that underpin it. The findings were largely to be implemented by 'community action' by the parties connected with the industry. Later reports such as Latham and Egan undertook 'research' – and the word is used approximately – to define targets for the industry that would be driven into the practice of the industry by a combination of client pressure, Government intervention by legislation (The Construction Act) or institutions established to oversee implementation. Consequently, the industry is constantly under surveillance. Work is benchmarked, KPI'd, CALIBRE'd and subject to other performance measurements. It could be that a consequence of such surveillance has been to diminish the pleasure that people can gain from work in the industry. The industry has consequently become less attractive, and the poor recruitment to the professions and trades could have long-term impacts. The demands being made on the industry cannot be met and so lead to an industry that cannot attract staff to deliver buildings on time, with increased costs and questionable quality.

1.5 References

1   National Audit Office (2001) *Modernising Construction.* Report by the Comptroller and Auditor General HC 87 Session 2000–2001: 11 January.
2   Simon, Sir Ernest (1944) *The Placing and Management of Contracts.* HMSO, London.
3   Construction Task Force (1998) *Rethinking Construction.* Report of the Construction Task Force to the Deputy Prime Minister, John Prescott, on the scope for improving the quality and efficiency of UK construction.
4   Butler, D. & Butler, G. (2000) *British Political Facts 1900–2000.* Macmillan, Basingstoke.
5   Cain, C. (2001) *A Guide to Best Practice in Construction Procurement.* Construction Best Practice Programme Guide.

Chapter 2

Placing and Management of Building Contracts: The Simon Committee Report (1944)

Patricia Hillebrandt

2.1 Background to the study: the economic, political and social climate

The committee that produced this report, under the chairmanship of Sir Ernest Simon, was set up in December 1942, only about half-way through World War II, and presented in May 1944, still a year before the end of war in Europe. It was therefore an example of quite remarkable forward thinking. Yet at the time of its publication the Government had already considered post-war training for the building industry[1] and post-war employment policy[2], had set post-war targets for house construction and construction industry manpower and would produce a Housing White Paper[3].

There were 41 members of the Committee, representative of all parties in the industry and including names still familiar nearly 60 years later, such as Mr H. Manzoni, later Sir Herbert, the Birmingham City Engineer and Surveyor, Mr J. W. Laing , later Sir John, contractor, and Mr J. W. Stephenson, later Sir John, trade unionist. The Chairman of the Committee had chaired the 1942 report on training and, as Lord Simon, would continue to serve Government in such endeavours.

It was important for the Committee to consider the environment in which change could take place. In considering the economic environment, it clearly had to consider the situation as it was in 1944 and, even more, as it believed it would be after the war was over. However, for the discussion of the matters relating to the placing and management of contracts, the situation in 1944 was less relevant because it would have been expected to be a transient situation only. In particular, contractual arrangements and payment methods for war damage repairs, which often had to be done before anyone had assessed the likely cost, was something peculiar to the situation of continuing damage by various forms of enemy action and would become irrelevant after the war was over. Similarly, rapid building of military installations would also be irrelevant.

Consequently, the Committee, in comparing the post-war situation with what had happened in the past, needed to look at the period prior to 1939 for, by 1939, the interest was focused on war preparations. Indeed, consideration had been given to the dangers to be faced in wartime for many years. *The National Builder* in March 1937 carried an article on 'Some principles of protection in air raids' which reported that, as early as May 1935, the Air Raid Department had been formed as part of the Home Office[4]. In the press prior to 1939 there was discussion of the current recession and also some articles on the organisation of building work and of contractual arrangements. A paper given by Oliver Roskill to the Royal Institute of British Architects (RIBA) in 1938 set out some of the problems of the industry that were inherent in the way the process worked[5]. It is clear that the construction process did not always proceed smoothly.

At the time the Simon report was published, the Government and industry were looking ahead to the end of hostilities. A browse through the building press in the war years shows how the industry, as well as Government, was concerned to plan for the post-war years. Even in the desperate year of 1942, the editorials of *The Builder* were forward looking and the articles similarly so. In February 1942 *The Builder* carried a series over 5 weeks on planning the post-war industry[6]. The National Federation of Building Trades Operatives (NFBTO) produced a pamphlet on post-war building policy in April 1942[7]. In the year in which the Committee reported, the emphasis was on review and change, covering subjects such as housing, the future of the architect, the organisation of building contracts, town planning and the future of the industry. *The National Builder* had similar concerns, although it devoted more attention to the day-to-day problems. Professional organisations were also busy. The RIBA, for example, appointed a committee on national planning[8]. Government was working hard on the future of the industry. The Minister of Works appointed an expert mission to study building methods in the USA and it reported in February 1944[9]. In January 1944 the Building Apprenticeship and Training Council, set up after the report on training for the building industry[1], published its first report[10].

The most urgent work was to continue the repair of war-damaged buildings that could still be used. The second priority was to deal with the great shortage of housing which was a major worry. *The National Builder* devoted considerable attention to housing, both long term and short term, and reported developments in the design and production of prefabricated housing as well as the overall needs at the end of the war. It also reported Churchill's very keen interest in the provision of housing[11]. In 'The Second World War, Closing the Ring'[12], Churchill referred, in April 1944, to the prefabricated house (although he disliked the name and suggested 'ready-made' instead[12, p. 618]) and asked that more be put on show to be seen by 'working women and people of all classes'[12, p. 625]. In September 1944, Churchill wrote a letter to Sir Edward Bridges, expressing his exasperation at the lack of progress. 'What does it matter,' he said, 'whether the house is the best thing that can be built or not? The great thing is to have some kind of a house for a soldier returning who wishes to marry'[13, p. 607]. He appointed a Cabinet

Committee under Lord Beaverbrook and asked for a plan of action in a fort-night's time[13, p. 607]. In March 1945 he was pushing for labour for the building industry immediately after the war, including the formation of special housing units and direction of some men having early release from the army[13, p. 632]. After the end of the war and before the general election, he was again concerning himself with the programme for housing construction and proposed to create five regiments of a thousand men each to get the programme under way in terms of emergency shelter and temporary and permanent houses. He was also concerned about the supply of labour and materials and about rent control[13, p. 652]. He did not have the opportunity to put his plans into practice because he was defeated at the 1945 election.

The Labour Party shared the view that controls were essential, and in their election manifesto[14] they pledged that:

> 'it [the Labour Government] will proceed with a housing programme with the maximum practical speed until every family in this island has a good standard of accommodation. That may well mean centralised purchasing and pooling of building materials and components by the state, together with price control.'

Thus, the atmosphere in which the Committee was deliberating was one of intense interest in improving the efficiency of the industry so that, after the end of the war, it could deliver the construction required and especially housing. All parties to the process, Government and Churchill himself were involved. In spite of this, the response to this report in the building press was negligible. It would seem now that the way contracts are arranged is an important factor affecting the performance of the industry. Perhaps the publication was overshadowed by other events at the time. Perhaps the very technical and legalistic nature of some of the arguments made the report uninteresting to the general building industry reader. The delay of some months in publication of the report after it had been presented to the Minister may have had an effect. The next section of this chapter describes the content of the report. The reader may find some clues there as to the reason for the low level of publicity it received.

2.2   Main concerns of the report, conclusions and recommendations

2.2.1   *Introduction*

The brief for the Committee was to examine the whole question of the placing and management of building contracts, to consider how far existing practices were suitable and to make recommendations ensuring an improvement in building organisation so as to provide the best possible service to the nation while maintaining an efficient and prosperous industry. The report deals mainly with the work of large and medium builders. Indeed the only recommendation concerning the small jobbing builder is that he should have a better technical training. Note that the terms of reference did not include civil engineering.

In its introductory section, the report notes the range of types of consultant and specialist firm in the industry and their high degree of technical and scientific knowledge. It draws attention to the vital function both of the architect and of the contractor to organise and coordinate the work of others. It comments also on the wide range of standards in the building industry and the difficulties of defining standards, let alone enforcing them, in an industry where the product is so diverse.

Site management, it notes, is extremely complicated and, compared with management in factory conditions, is incomparably more difficult in respect of the use of plant and machinery, supervision, the conditions of work and welfare, owing to the inherent characteristics of the construction industry. The report authors acknowledge the problems of the contracting firm, notably the fluctuations in the volume of work on its order books at any one time and the consequential variation in the numbers of men employed. In any case, because of the geographical dispersion of work, the contractor can keep only a small nucleus of permanent staff who must be prepared to travel.

The report stresses how important it is to provide conditions that will facilitate and encourage:

- the selection of men and firms according to their character and ability;
- responsibility and pride in work;
- fair remuneration for good service.

Under these conditions the owner should have confidence in the integrity, ability and goodwill of all parties concerned in the contract. It is the view of the authors of the report that these conditions are fundamental to a successful building industry.

The main body of the report, which was intended to be suitable for laymen, considered, in eight chapters, aspects of the industry and made recommendations for improvement. A review is given here of the main arguments of the report and the resulting recommendations, taking some licence to give more weight to those matters of interest at the start of the twenty-first century. This review will keep the same subject headings as these chapters, except that the chapter on 'The Economic Background', which is the last chapter in the report, is here dealt with first, because it illustrates the environment in which the report was prepared. In addition, the report published seven technical papers which were, to a large extent, the basis for the conclusions and recommendations of the report. Relevant points made in these papers are incorporated under the main chapter headings of the report.

2.2.2   *The economic background (Chapter 9 of the report)*

The chapter on the economy is supported by and, to a large extent, based on a report by an anonymous economist. The ground covered in this report is wider than that in the chapter because some of the comment and recommendations are outside the scope of the report.

In the period between the two World Wars there was a recurring problem of booms and slumps with the very serious unemployment that characterised the industry during this time. The worst problem, as seen by the Committee, was irregularity in the total orders put to the industry, in the orders of individual firms and in the employment of operatives.

There were few statistics of the industry prior to World War II, and the report by the economist concentrates attention on housing which was a very important component of total output at that time. From 1920 the number of houses built rose only slowly from around 70 000 and did not reach nearly 270 000 until 1928. It then fluctuated between about 200 000 and 370 000. It is not surprising that the industry was concerned at the irregular workload. There was general disappointment that, in spite of determination to deal with the serious shortage of housing after World World I, reflected in a scheme known as the Addison scheme, and, in spite of lavish subsidies (partly due to the need to control rents), the rate of building was so slow. The economist's report identifies the main reasons as pressure to increase the rate of building in spite of shortages of materials and labour, unlimited subsidy and no control over non-essential work. The material shortage was slowly overcome, but the trade unions were unwilling to sanction as many new entrants to the industry as were needed because they were afraid the boom conditions would not last, leading to unemployment. At the same time, profits were large.

The Committee was anxious to learn the lessons of World War I, at the same time realising that the objectives were not identical. Some objectives for the post-war period had already been defined by Government and included the following:

- in the words of the report, 'The fundamental aim of the Government must be to secure as nearly as possible full employment for the operatives and a regular flow of orders for the contractor';
- four million houses to be built in the 10 years after victory is achieved;
- the workers in the building industry to be brought up to one and a quarter million in 3 or 4 years.

To meet these objectives, the report states that it as essential to formulate a building programme and to regulate the volume of building in line with the programme. At the same time a manpower plan is also required to assess the amount of manpower needed to meet the building programme and to proceed, well in advance, to provide the necessary manpower. It is noted that adjustment of the total manpower must be slow because it takes 5 years to train a craftsman. In addition, a plan for materials is also necessary to ensure that materials are available in the right quantities and at the appropriate prices. In these matters the Committee was satisfied that there was considerable experience in the Ministry of Works from the work they had had to do during the war.

There was concern that trade associations had been fixing prices for the last 15–20 years and that, unless they were controlled, they could fix prices at a high level. The Committee recommended the registration of all price-fixing associations and that the Ministry of Works should have full information on their activities and power to take whatever action was necessary.

In conclusion, the report states that:

'the success of the building programme in the post-war period depends on effective action by the Government to regularise the demands on the industry, to stabilise employment and to prevent unreasonable prices for building work.'

2.2.3 *Precontract preparation – the time and progress schedule (Chapter 1 of the report)*

This chapter starts by describing the traditional process in construction, with the architect leading the team on behalf of the client. This system is in general regarded as the right system with inefficiencies traced to:

- insufficient precontract preparation;
- too many variation orders;
- indiscriminate competition leading to work being given to builders with lowest standards;
- indefinite relation between contractor and nominated subcontractors.

Much of the responsibility for things going wrong is placed on the shoulders of the building owner for not choosing his architect well, not keeping in touch with and supporting him and, above all, chivvying him to advance each stage of the process, for example, design and tender documentation, with inadequate preparation.

It is necessary not only to establish what has to be done but also to time the operations so that the tasks of all the different parties fit in and do not interfere with each other. For this, a time and progress schedule is recommended. This should be part of the main contract and should include:

- the start and completion dates of every stage of the main contract and of subcontracts;
- the date of handing to the contractor the necessary additional detail drawings.

The date for drawings had not previously been included in the schedule. For each item progress should be monitored. An advantage of the schedule was that it would be possible to establish which party was at fault in not meeting time deadlines so that damages could be assessed if necessary.

2.2.4   *Types of contract (Chapter 2 of the report)*

The report distinguishes two main types of contract: fixed price and cost reimbursement contracts. Several varieties of fixed price contracts are described:

- lump sum with no bill of quantities;
- lump sum based on a bill of quantities;
- schedule contract in which the sum paid to the contractor is determined after the work is completed by applying prices in the pre-prepared list of prices to the quantities of the work actually done;
- rise and fall clauses in which the price is varied with the changes, over the contract period, in prices of inputs.

The latter is not truly a fixed price contract but had been used after World War I and during World War II when there was great uncertainty on the future prices of materials and various labour costs.

The cost reimbursement contracts are described as:

- cost plus percentage contracts where the contractor's overheads and profits are covered by a percentage added on to costs incurred;
- cost and fixed fee contracts where the overheads and profits are covered by a predetermined fixed amount;
- value cost contracts where the price is calculated according to an agreed schedule of prices and the fee rises if the final cost is less, and vice versa.

Fixed price contracts are strongly recommended. If a cost reimbursement basis is used the fee should be a fixed sum. Value cost contracts are suitable only for large, continuing, experienced clients.

Direct labour departments were set up between the wars mainly to build houses. The report lists the conditions on which the success of direct labour departments is seen to depend:

- competent managerial staff and director;
- as much freedom as a private contractor for the director to deal with staff and labour and material purchases;
- work to be confined to repetition jobs, almost entirely housing;
- maintenance of a steady workload;
- an independent costing department to ensure that all costs, including overheads, are debited to the job to enable comparison with contractors' prices.

2.2.5 *The selection of the general contractor (Chapter 3 of the report)*

The traditional way of selecting the contractor is by competitive tender and to award the contract to the lowest tenderer unless there is some reason to pass him over. This can be subject to abuse by permitting contractors who are not in a position to carry out the work to the required quality to tender, and also by allowing too many tenderers, resulting in too high a cost to the industry as a whole. The Committee makes strong recommendations on this matter. It states:

> 'Competitive tenders should in all cases be called from a limited number of firms carefully selected as being capable of and likely to do work of the required standard.'

There is no difficulty in the private client doing this with the help of his architect, but there are very substantial difficulties for local authorities.

Under the joint effect of the Local Government Act 1933 and the Model Standing Orders issued by the Ministry of Health, local authorities must advertise tenders and accept the lowest tender no matter from whom it is received. The Committee regards this practice as being 'contrary to the interests both of the building owner and the building industry' and recommends that it should be abolished. The report acknowledges that the practice arises from a desire to avoid any suggestion that the local authority is not acting impartially. It states that, in view of the damage being done by the present system, another way must be found to avoid the risk of giving local authorities discretion in these matters.

Government departments have acted in the same way, but during the war they tended to invite contractors from selected lists recommended by their advisers, and these lists, kept up to date, could provide the basis of a list from which to draw tenderers:

> 'Government departments should in no circumstances revert to the habit of issuing indiscriminate permission to tender.'

An alternative to tendering is negotiation with a selected contractor. This has the advantage that the contractor can be selected early and is then available for consultation on questions of construction and selection of subcontractors. Using this method, the owner may not be paying the market price for his building. Large building owners with much work find this procedure satisfactory.

2.2.6 *Selection of specialist subcontractors (Chapter 4 of the report)*

Subcontracting is desirable for the efficiency of the industry. In general, the specialist trades should be subcontractors to the main contractor. However, there are some trades that should be nominated by the architect. These include those trades for which the design has to be completed by the specialists before the main drawings are complete, such as structural steelwork and heating and ventilating plant which are an integral part of the building. These nominated subcontractors should be selected by tender, with design fees being separate. Work where the architect wants a certain subcontractor because of his particular skills may also justify nomination. Difficulties can arise in the case of bankruptcy of the main contractor or the nominated subcontractor.

2.2.7   *Bills of quantities (Chapter 5 of the report)*

The report regards bills of quantities as the best basis for estimating the cost of the whole project and for valuing any variations.

The Committee was concerned at the detailed work in setting out and valuing many small provisional items which form a large proportion of the total items but a small proportion of the total value of the work. It accordingly recommends that, for small items, a trade schedule should be provided by the industry, in collaboration with the Ministry of Works. In pricing, the contractor would quote a percentage on or off. This would save time and labour for the quantity surveyor and for the contractor. It would also reduce the problems of a large number of small items.

The report states that it would be desirable to have abbreviated bills of quantities for small housing projects while prices are unstable.

2.2.8   *Site management (Chapter 6 of the report)*

The report emphasises the importance of all participants in the construction process having a clear idea of their tasks and their timing, as should be spelt out in the time and progress schedule. The architect is in control until the contract is signed and later for seeing that the contractor carries out the work in accordance with the contract and according to his reasonable satisfaction. The architect should give all instructions to the main contractor who should deal with the subcontractors as appropriate. Any variations should be given in writing. The detailed drawings should conform to the bill of quantities or, if this is not practicable, the architect must put his new instructions in writing. All variations should be discussed at the next site meeting.

Site meetings should be held regularly in an organised way with all relevant parties present. The architect should be Chairman and the contractor Deputy Chairman.

At present the Clerk of Works, who represents the architect on site, has no legal status. Especially on larger jobs, the Clerk of Works should have delegated powers to make decisions on day-to-day matters. To assist in this role, he should be given better technical training.

Because of the technical complexity and the increased number of subcontractors, the management of a large contract has become so difficult and so important that the contractors should specialise more and more on the science and practice of management. They should have more well-trained experienced managerial staff.

The contractor has a discount of 2.5% on subcontractors' work. It is recommended that this be discontinued. It implies that the main contractor has few responsibilities for the work of the subcontractor, but this is not so.

2.2.9   Personnel management (Chapter 7 of the report)

The report starts this section as follows:

'The public and Parliament have made up their minds that a policy of full employment should be a first objective throughout the whole of industry. This means a fundamental change of the conditions of work in the building industry; the risk of loss of work and of income will have gone for the competent honest workers.'

It recommends that, in conjunction with the industry, labour exchanges take special steps to reduce the intervals for the operatives between successive jobs. Furthermore, a special officer should be appointed at labour exchanges to deal solely with building labour.

The report also states that:

'during the war there has been an almost revolutionary improvement in the conditions of welfare in the building industry.'

It recommends that:

'they should where practicable be continued after the war by joint agreement between employers and trade unions.'

The report stresses the importance of good personnel management, pointing to the selection and education of personnel officers as one of the aspects of the building industry where there is most room for improvement. It recommends that more attention be paid to the training and selection of managers from the agent to the chargehand, with special attention to the difficult problem of personnel management.

The report observes that there has been a widespread increase in works committees in industry generally. On large construction jobs there should be a works committee to secure proper cooperation between management and operatives. Moreover, workers should be given information about the progress of the job they are working on:

> 'The Ministry of Works, in conjunction with the industry, should organise a national information service to help the individual contractor.'

2.3   The impact of the report

2.3.1   *The economic background*

Although the subject of the report was the placing and management of building contracts, the Committee makes strong recommendations on the management of the demand on the industry and the way demand and supply should be matched. It advocates a plan for demand, for materials and for manpower.

On the overall management of the demand for building, the report recommendations were substantially accepted. In stark contrast to the situation after World War I, building licensing for private work continued after the war until 1954, although it became easier and easier to obtain licences as conditions improved. Work in the public sector had to be authorised by the appropriate Government department. The control operated partly through the laying down of a starting date, first for private work and later also for Government work[15]. In the immediate post-war years the controls were not applied vigorously or efficiently. More building work was licensed than was done. In part the policy was to allow housing to compete for resources with repair[15]. This was one of the things the Simon Committee wanted to prevent. In addition, the supply of building materials was a constraint.

The building materials that were rationed were steel and timber, but initially there were severe shortages of unrationed materials, especially bricks[15]. It appears that the Simon Committee envisaged a much stronger allocation procedure for building materials than actually took place. By 1947 the industry was badly overloaded, partly because of the unexpected shortage of steel. Stronger controls over demand were introduced. This change in policy led, rather unexpectedly, to surpluses of other materials such as bricks and plasterboard[16].

Government had concerned itself, well before the end of the war, with ways of rapidly increasing the building labour force after the end of hostilities. The Ministry of Labour compiled a register of all persons who had worked in the construction industry and were then employed elsewhere and tried to persuade them back into the industry. Building workers who agreed to return to the industry were given priority release from the armed forces.

Government also established training centres. The schemes worked well. From June 1945 to the end of the year, the labour force of the building and construction industry increased from 520 000 to 682 000, and by the end of 1946 it was 953 000. By 1954 it was 1.2 million compared with under 0.8 million in 1935[17, pp. 42–43]. The path to this increase in numbers was not always smooth. The unions accepted the national apprenticeship scheme but were against the 6 month training of adults. They were afraid of dilution in the industry and the danger of future unemployment when demand decreased. However, agreement was ultimately reached[18, p. 208].

2.3.2   Time and progress schedule

It was necessary for the Committee to make the point that there should be such a plan of work. A memo dated 11 June 1943 was sent by Godfrey Mitchell[19], the Chairman and founder of George Wimpey in its modern form, to a certain W. H. T., in which he said:

'When I look back on our house-building days, we never did plan the progress of the works; we simply let the navvy go ahead as fast as he liked, the concretor went along behind him, and then the bricklayer, so that very soon we got the whole 100 houses up, with the whole of the labour force which remained on the job being spread throughout the whole 100 houses – you never knew where the plumbers were, where the electricians were, and nothing was ever finished.'

This statement was probably an exaggeration for effect. Many house builders had since prewar days used an alternative system of starting houses one by one or in small batches and working on them so that finished houses became available in sequence and the resources locked up in work under construction were minimised. By 1945, Wimpey were operating their sites so that completions were in line with demand (E. J. O'Neil personal communication). The point Godfrey Mitchell was making was that improvement was necessary and he went on in the memo quoted above to outline how this situation could be improved and to discuss the likely type of construction after the war[19].

Similarly, good pre-war site agents were using some form of planning for their larger jobs, and certainly by 1942 basic bar charts were in use. Indeed, by November 1942, time and progress schedules had been used on all work under the charge of the Ministry of Works and Planning[20]. This did not apply to war damage work where contractors were paid on various versions of cost plus. Indeed, the direction to builders to send so many men to a specified site immediately to carry out urgent repairs, though necessary, killed all planning of workload or sequence of operations.

For new work the situation did improve. Probably, the practices of the most efficient builders became general. A schedule in some form or other, often a bar chart, was being used generally for large projects and by large contractors for small projects by the mid-1950s. Critical path programming, as used by NASA, was introduced from the USA, with Richard Costain Ltd to the fore, in the late 1950s (E. J. O'Neil personal communication).

However, the idea put forward in the report that the time and progress schedule with full detail should be part of the contract has never been implemented. Moreover, the architect does not normally produce any detailed schedule. The contract provides for dates of starting and completion and sometimes dates for the completion of certain stages. The contractor's bar chart would not normally be available to all parties, and the critical path programme never so (E. J. O'Neil personal communication).

The report emphasises that the timing of detailed drawings should be included in the schedule. It does not discuss whether detailed drawings should ideally be completed before the tender is let, but it strongly warns against the client proceeding too soon to letting the work to a builder before the details are thought out. If all drawings were available at the time of tendering, the contractor would be able to make a better estimate of final cost and would have much less opportunity to claim on clients. The architect's fee payments make provision for a large proportion of the fee being paid at or before tender stage. The architect does not therefore have a vested interest in delaying the production of the drawings. An important reason why drawings are not done earlier is the pressure from clients for speed in commencing work on site and indeed to shorten the overall time of the total process. Moreover, it is neither in the contractor's nor in the quantity surveyor's financial interest to have a completely predetermined project because claims for lack of drawings and variations are profitable to them. The Committee seems to accept that it is acceptable for the detailed drawings to be available at intervals through the construction process. It was perhaps facing the hard reality of the process.

2.3.3   Types of contract and the selection of the main contractor

Immediately after the war, when there was great uncertainty in prices and availability of supplies, the cost plus contract was in frequent use, especially for repair and maintenance work. It was also in use because new techniques of construction with different materials and a larger prefabricated component were being introduced, which meant that there was not enough experience to give a reliable estimate. Negotiated contracts were more widely used for the same reasons and brought the building contractor in at an earlier stage in the process so that he could contribute to the pool of expert knowledge[21, p. 415]. This, however, was the exception in local authority work.

In general, local authorities regarded open tendering as the norm. Bowley[21, p. 415] states that:

'the Ministry of Housing and Local Government had actually discouraged any modification of traditional contract procedures. As recently as 1957 new Model Standing Orders issued to local authorities in England and Wales confirmed the general rule of open public invitations to tender for building contracts. The explanation issued with the Model Orders pointed out that the Orders had been drawn up so as to make it clear that abrogation of the rule would be exceptional.'

According to a survey by The National Economic Development Office (NEDO), quoted in the Wood report, even in the 1970s 16% of public sector clients used open competition for the selection of main contractors for building projects over £50 000[22]. In spite of the abolition of open tendering being one of the main recommendations of the Simon report, unfortunately it had little impact on practice for many years to come. Only very slowly did local authorities abandon the practice, helped by the building up of approved lists of tenderers.

As open tendering was slowly replaced by selective tendering, another problem arose. The numbers of contractors invited to tender was far too high – often more than 12 – and the costs of obtaining a successful bid continued to be unnecessarily high. It was decades before the National Joint Consultative Committee (NJCC) recommended sensible numbers[23].

Meanwhile, the private sector was experimenting not only with selective tendering but also with package deals or turnkey projects which were being developed by the late 1950s, something not envisaged in the Simon report.

The conditions of success for direct labour departments laid down in the report were never realised, and there was constant complaint about the way the direct labour departments functioned until the time of their effective demise.

2.3.4   Site management

It is still necessary to draw attention to the need for an orderly chain of command and for efficient unambiguous practices to be observed. Since the Simon report there has been a great improvement in the training of managers for the construction industry. There are now many degree courses that provide a good background for site management, although they do not replace the need for experience and training on the job. The great advantage of the site agent who had been promoted from the trades and then from various site supervisory jobs was that the selection process ensured that he was a good manager of men. Furthermore, at that time the education process had no mechanism to select the brightest students for further education, so that many might find their way into the building trades. The improvement in education in building management has been assisted not only by degree courses in universities and former polytechnics but also by The Chartered Institute of Building. This organisation developed, over a period of many years, from the Institute of Builders which, at the time of the report, was virtually moribund[24].

One of the matters mentioned in the report as needing change is the 2.5% cash discount received by contractors on subcontractors' work. The Committee thought that the discount diminished the perception of the importance of the main contractor's supervision of the subcontractor's work. As the main contractor is responsible for the subcontractor's work, whether he is a domestic contractor or a nominated contractor, this argument is difficult to accept. If this discount by subcontractors to main contractors were in reality a 'reward' for prompt payment, then it would have some merit. It is understood, however, that now it is often taken off by contractors at the time they pay their subcontractors as a matter of course, whether payment is prompt or not. It then becomes an extra mark-up for the main contractor on the subcontracted work in the contract. If, however, it is standard practice, then it will be allowed for in estimations both by the main contractor and the subcontractor and there will be no net effect. It does not seem a major factor affecting the industry's efficiency. The recommendation in the Simon report to discontinue it has certainly not been heeded.

2.3.5   Personnel management

Conditions on building sites have improved since the Simon report. There are now minimum requirements for facilities for site employees, although many were not introduced until the 1960s when the Ministry of Public Building and Works, as it had become, reviewed the situation. Conditions are still unsatisfactory compared with other industries. However, with very competitive conditions in the industry, no contractor can afford to improve site conditions unilaterally. To achieve improvement, better conditions must be included in the contract.

The problem of short-term casual employment of labour by contractors was never resolved. The casual employment of direct employees has now been replaced by labour-only subcontracting under which the workers are self-employed. This does nothing to remove the hardship when the workload on the industry declines and is, in many ways, worse from the point of view of security because the self-employed forfeit some benefits of the directly employed, notably unemployment pay.

In this section, too, the importance of training is mentioned with special reference to personnel officers.

2.4   Conclusions

It is difficult, even impossible, to say what would have been different without the report. The industry hardly took notice of the report when it was published. Government probably did take notice. In that part of the report dealing with the economic background, the Committee was to some extent exceeding its terms of reference. Whether the Government was influenced by the report's recommendations on plans for level of demand, materials and labour is not known. The fact is that it did embark on what was effectively the first attempt at national planning, in this case trying to ensure a balance between the building programme and the resources available to undertake it. The recommendations of the Simon report came at a propitious moment in that they reinforced the current thinking of Churchill on these matters and may well have carried weight with the next Government.

As regards the main subject of the report – the placing and management of contracts – its success is limited in terms of implemented recommendations. Indeed, in some areas so little has changed that some of the report could be written today. Yet the report has been quoted again and again and is considered by students of the construction industry as one of the milestones in the literature of the industry on the management of the construction process. It was the first major broad-based report commissioned by Government into the way projects were procured by clients.

The next major landmark in this respect was the Emmerson report[25], which is the subject of Chapter 4. In between Simon and Emmerson, Government was the instigator of a number of other reports on the construction industry. The Allen report[26] had a broad remit to look at the organisation and efficiency of the industry and, after a description of the industry, dealt with such subjects as productivity, contracts, planning of projects, technology and finance. The Phillips report[27] also has significance and is the subject of the next chapter.

There were many reports on housing, still the preoccupation of Government. A general report already referred to[3] and two on the costs of housebuilding[28, 29] are examples of many. Stone describes in some detail the development of thinking on non-traditional building methods during and after the war, including the activities of an interdepartmental committee, originally set up in 1942 but active for many years[30]. Building materials were the subject of scrutiny in 1948[31]. The Building Research Station was at that time concerned with economic aspects of the industry, especially housing, as well as with the way the construction process worked. To a certain extent, Simon was a forerunner of these investigations.

The greatest achievement of the Simon report was probably to help prevent a total mismatch of supply and demand for the services of the industry and the consequent chaos in the industry that could have occurred after World War II without such foresight. The second achievement was to open up the subject of the way the construction process should be organised to serious, comprehensive debate. The relatively minor changes made to the detail of the placing and management of contracts were probably less important.

2.5   References

1   Ministry of Works and Planning (1942) *Report on Training for the Building Industry*. Central Council for Works and Buildings, Education Committee (Chairman Sir Ernest Simon). HMSO, London.
2   Ministry of Reconstruction (1944) *Employment Policy*. Command 6527. HMSO, London.
3   Ministry of Reconstruction (1945) *Housing*. Command 6609. HMSO, London.
4   *The National Builder* (1937) Some principles of protection in air raids. *The National Builder*, March, 274–5.
5   Roskill, O. W. (1938) Economics of the building industry. *The National Builder*, December.
6   *The Builder* (1942) Planning the post-war building industry. *The Builder*, 20 February– 20 March.
7   NFBTO (1942) *Post-war Building Policy*. National Federation of Building Trades Operatives, London.
8   *The Builder* (1944) The RIBA and national planning. *The Builder*, 28 January.
9   Ministry of Works (1944) *Methods of Building in the USA*. HMSO, London.
10   Building Apprenticeship and Training Council (1944) First Report. HMSO, London.
11   *The National Builder* (1944) Mr Churchill and building problems. *The National Builder*, December, 93–4.
12   Churchill, W. (1952) *The Second World War, Closing the Ring*, Vol. 5. Cassell, London.
13   Churchill, W. (1954) *The Second World War, Triumph and Tragedy*, Vol. 6. Cassell, London.
14   Labour Party (1945) *Let Us Face the Future*. Labour Party, London.
15   Dow, J. C. R. (1965) *The Management of the British Economy*. Cambridge University Press, Cambridge.
16   Bowley, M. (1960) *Innovations in Building Materials*. Duckworth, London.
17   Rosenberg, N. (1960) *Economic Planning in the British Building Industry 1945–49*. University of Pennsylvania Press, Philadelphia.
18   Kingsford, P. W. (1973) *Builders and Building Workers*. Edward Arnold, London.
19   Mitchell, G. (1943) Abstract of a copy memo forming part of a file that belonged to Godfrey Mitchell, held at Kimmins Mill, Stroud, Gloucestershire on behalf of the Construction Industry Resource Centre Archive (CIRCA), managed by WICCAD.
20   *The Builder* (1942) Time and progress schedule. *The Builder*, 6 November.
21   Bowley, M. (1966) *The British Building Industry: Four Studies in Response and Resistance to Change*. Cambridge University Press, Cambridge.
22   EDCs for Building and Civil Engineering (1975) *The Public Client and the Construction Industries (the Wood report)*. NEDO. HMSO, London.

23   NJCC (1972) *NJCC Code of Procedure for Selective Tendering*. National Joint Consultative Committee, London.
24   Hillebrandt, P. M. (1984) *Analysis of the British Construction Industry*. Macmillan, London.
25   Ministry of Works (1962) *Survey of Problems before the Construction Industries* (Chairman Sir Harold Emmerson). HMSO, London.
26   Ministry of Works (1950) *Building*. Working Party Report (Chairman Professor G. C. Allen), HMSO, London.
27   Phillips Report on Building (1950) Working Party Report to the Minister of Works. HMSO, London.
28   Committee of Inquiry appointed by the Minister of Health (1948) *The Cost of Housebuilding, 1st Report*. HMSO, London.
29   Committee of Inquiry appointed by the Minister of Health (1950) *The Cost of Housebuilding, 2nd Report*. HMSO, London.
30   Stone, P. A. (1983) *Building Economy*. 3rd edn. Pergamon, Oxford.
31   Ministry of Works (1948) *The Distribution of Building Materials and Components*. HMSO, London.

Chapter 3

The Working Party Report to the Minister of Works: *The Phillips Report on Building* (1948–50)

Alan Wild

3.1 Methodology, approach to the study and preassumptions

The Phillips report[1] is treated contextually[2] with a historical span between Simon[3] and the National Economic Development Office (NEDO) report on construction[4]. Construction reports are contexts for each other, hence:

> 'the truth theory has to be qualitative confirmation since the context will change and knowledge will also need to change, and the root metaphor is the historical event.'

The chapter develops a previous conference paper[5] to confirm the argument of Higgin & Jessop[6] that post-war reports:

> 'have not been able to do more than canvass best opinion and agree general precepts on that basis'

as they have lacked an analytical framework for the functioning of construction operations in their wider context. The writer agrees with Andrews & Darbyshire[7]:

> 'If so many have preached the unification of the UK construction industry for so long with so little effect how could we succeed? Is it indeed a lost cause? Have we been barking up the wrong tree all these years? Is it our destiny to build a fragmented future on a fragmented past and make the best of it?'

The Phillips report reveals early on the extent of this fragmentation, establishing itself as a template for later reports, especially in the themes and problems they discuss and their putative solutions. Hence, there is a continuity within the sequence, a conventional wisdom[8] revealing 'the ideas in good currency'[9] from particular historical contexts transported in as solutions to currently assumed 'problems of construction'. Both problems and solutions are recycled in different forms revealing construction's recalcitrance[10] and raising the question as to whether the problems of construction yield to rational inquiry at all?

3.1.1 Terms of reference and the composition of the Working Party

The report initiated in July 1948 is the first of that long sequence of official inquiries that have punctuated construction's post-war history. While Simon[3] focuses on 'contracts', Phillips adds 'coordination' and this in turn moves on to 'communications', the other strand of the discourse of post-war reports. Phillips' terms of reference are:

> 'To inquire into (a) the organisation and efficiency of building operations in this country, including those of the specialist and subcontracting trades; (b) the position of the professions in relation thereto; (c) the arrangements for financing operations; and (d) the types of contract in general use and to make recommendations.'

Latham[11, 12] inquired alone, with supporting assessors from the metropolitan professions and large construction firms. The Working Party contained four national trade union officials and received evidence from the Communist Party and The Cromer Street/Kings Cross Works Committee, presumably a joint production committee of a direct labour organisation. Post-war full employment and the enhanced political standing of the unions insulated pay from the direct terms of reference. However, the trade union leaders had metropolitan status, so Phillips continues the wartime consensus between trade union officials, industrialists and academics:[13]

> 'a common doctrine and outlook, a readiness to use the same methods and to move towards the same conclusions, and a will to cooperate.'

This value consensus around shared experiences and collaboration across the public and private sectors is a form of 'dirigisme' that I label without discussion 'Keynesian planning'. In this report, such 'planning' addressed supply-side issues.

One member, Sir Hugh Beaver, exemplifies this consensus, its continuity into the 1960s and, in construction, its methodology of arm's length influence through networks. He was wartime Controller General of the Ministry of Works and later Managing Director of Guinness. In the 1950s he participated in management education initiatives for construction. As Chairman of the Tavistock Institute of Human Relations (TIHR)[6] at the time of its selection to conduct a radical analysis of construction, he personified TIHR's quasi-corporatist standing: safe hands for an innovatory project for which he became a trustee[14]. Phillips' influence terminates as Beaver retires from active involvement in construction.

3.1.2   *PEST (Political Economical Social Technological) analysis*

Reports on construction are a part of the industry's context. Phillips followed Simon[3] fairly rapidly and accompanied the other Simon report *The Distribution of Building Materials and Components*[15] which again reflected a preoccupation with supply-side problems. Post-war aspirations and plans for reconstruction were reflected in the 1944 Simon report[3] and the climate of 'planning for post-war Britain'[16] as means to desirable social objectives with the state as a provider and promoter of significantly enhanced power and authority crucial for both direct orders and macroeconomic regulation[17]. Phillips reports at the upturn of the long building cycle of the 1950s to 1980s. Output in real terms rose rapidly until the mid-1960s. In 1948, building and civil engineering accounted for 58% of gross fixed investment. Consequently, construction may be seen as creating its own economic context, especially in its readjustments to post-war conditions. Emmerson[18] comments on

> 'the remarkable recovery of the building materials industries and the construction industries from the war period when they were practically closed down; their flexibility in meeting new demands on their services in the past fifteen years; the introduction of new materials, increased mechanisation and new techniques; the steady rise in output, and the avoidance of major industrial disputes.'

The social context is a 'new Britain' of factories, hospitals, houses and schools: construction was a physical manifestation of popular expectations over education, employment, health and housing. The report reflects the acceptability of new values and concepts of technical progress as means and ends for the industry and its processes. Technological change is revealed in the changing composition of skills.

3.2   Contents

Phillips describes the effects of war on construction and its efficiency from 1939 to 1949, the state of building and of its participant organisations and manpower focusing on operatives who had experienced the largest degree of disruption. Future aspirations relate past trends and educated guesses as to what should and should not be done as a consequence of the slow recovery in productivity post-war. The report deals with a range of familiar problems:

labour productivity and operations; materials and methods; contracts; the professions and management; public procurement and planning regulations; research; finance and comparisons with the USA, Holland and Sweden. The appendices contain discussions of cost accounting and pre-war labour conditions.

Between 1939 and 1946/47, productivity fell by two-thirds, and by the end of 1948 it had fallen by three-quarters. Combined with higher wages and material costs, building cost inflation raised prices by a factor of 2.5 compared with their 1939 level. Labour cost rises of 2.8 were one-third of the increase and material cost rises of 2.2 were almost half the increase; profits and overheads rose almost by a factor of 3 compared with the pre-war level, amounting to 20% of the overall increase:

> 'in order to keep the matter in due perspective it should be remembered that they are not out of line with those which have occurred in industry generally.'

The labour force fell and increased rapidly, changing significantly in skill composition, management suffered dislocation, shortages and controls induced delays and uncertainties and the overoptimistic building programme was subsequently revised. Efficiency was affected by abolition by full employment of the previous reserve of unemployed labour and inducement of uncertainty and delays by planning controls, creating difficulties for preplanning of work. Temporary but important causes included the compulsory use of untried materials and the inflationary condition of the economy. These were improving.

Regulation and registration arrangements in relation to plumbing and electrical trades, the National Home Builders Registration Council(NHBRC) scheme and direct labour were confirmed. Prescriptions for improving efficiency included greater collaboration, preplanning of work, new technology, scientific management and the efficient organisation and dissemination of research. The structure of the industry provided flexibility and the range of capacities required, but greater craft flexibility would help. All categories of training, including that of architects in practical aspects of building, needed improving. Technical advice for public procurement was to be centralised in one Government department given the size, scope and complexity of public sector requirements.

Full working drawings for take-off of bills of quantities were to be provided before contracts were let with public procurement on a standard contract. Nominated subcontractors were not to proliferate and compete for a subcontract predesigned by a consultant. Full advance knowledge of the operation, a programme of work and careful site planning were required for management of building operations including supply of tools and materials and correct balance of operations and all other aspects of construction. Programmes and work progress were to be compared periodically and necessary adjustments made. Work studies could assist management by reduction in wasted time, and incentive pay schemes were essential if output was to be adequately increased. Joint production committees were seen as valuable, and adequate safety and welfare facilities were an important means of securing production. Appropriate costing systems were essential. Prefabrication of interiors and sustained support for standardisation would assist productivity. Careful use of new materials, more generally economical use of materials and extended use of power tools and mechanisation were essential for an adequate increase in output.

3.3   Themes and weaknesses of the report

3.3.1   *Segmentation, certainty and manageability as weaknesses of the report*

Problems are segmented between management and employers; designers and consultants; Government and operatives. The reluctance to acknowledge interpenetration of problems reflects underlying aspirations for manageability creating a search for certainty which conflicts with evidence. Assumptions of optimisation and improvement through formal rationality and functional methods, scientific management and rational costing proliferate. Uncertainties and disturbances surrounding procurement require rationalisation of contract arrangements and standardisation of controls and planning regulations. Collaboration and flexibility are preconditions for optimisation and improvement[5]:

> 'Collaboration is invoked through appeal to a virtuous circle of collaboration and rationality. This is a discourse of manageability.'

3.3.2   *Contracts*

Simon[3] is endorsed in relation to contracts, but there should be a standard contract for public procurement. The Royal Institute of British Architects (RIBA) contract is adequate and minor amendments unnecessary. The Joint Contracts Tribunal (JCT) should reconsider as too low the National Federation of Building Trade Employers (NFBTE) threshold of £1500 on competitive tendering for contracts through bills of quantities. Improved costing systems would establish certainty and accountability to the client and future cost planning for the construction process, aided by full working drawings for contractors to derive bills of quantities. Nominated subcontractors should be appointed after competition for a subcontract predesigned by a consultant but let by the main contractor.

3.3.3 *Conservative treatment of the professions and coordinating bodies*

The role and legal status of professionals and consultants, including architects' legal monopoly on design from 1931, is unchallenged. Stability of professional values and motives is assumed and treated as unproblematic. Spanning of boundaries and collaborative education should occur, but for future professionals, in spite of current problems of fragmentation. The designer's leadership role is unchallenged and coordinating bodies are deemed adequate for their purpose.

3.3.4 *Clients*

Clients are discussed from time to time. Strong clients could recover lost productivity:

> 'London County Council have claimed that on certain of their Value Cost Contract sites where incentive schemes are operating, they have regained their 1939 level of productivity, but this experience is exceptional.'

Uncertainty in clients and 'lack of decision and failure to supply fully informative drawings and particulars are increasing'. This had induced changes and variations, adding to cost and inefficiency and illustrating the advantages of complete architectural planning on the basis of clarity of the client with complete drawings available to the builder.

The scope and scale of building work for public clients combined the functions of owner and architect quite differently from ordinary building owners. Clear leadership and coordination of diverse approaches to construction were required, including dissemination of the implications of research for public procurement. The wartime Ministry of Works assisted this:

> 'but in our opinion the concentration of these functions has been allowed to fall short of the point at which full advantage would be reaped.'

Technical advice should be centralised.

3.3.5 *Operatives*

Pre- and post-war employment figures show that the workforce halved mostly owing to conscription. Between 1939 and 1948, civil engineering and building lost 230 000 operatives, including 63 000 semi-skilled: a key constraint postwar. Skill composition shifted. Specialist firms remained numerically about the same, but from 1945 to 1949 electrical contractors doubled, indicating the role of electrification in post-war modernisation. The report bypasses the exclusion of pay from its terms of reference by recommending strongly the introduction of incentive schemes to remedy the fall in productivity and endorsing improved welfare and site facilities as incentives. Pre-war indeterminacies were off-loaded on to labour where high levels of elasticity of supply guaranteed flexibility:

> 'a high margin of unemployment provided both a means of solving the organisational problems of the industry and also a disciplinary sanction.'

This obviated any need for costly managerial planning systems which remained to be developed post-war.

Post-war operatives faced excessive demands in an industry adopting new materials, components and methods and evolving technically. Radicalised by conscription, their new expectations were wished away as an attitude problem of a minority:

> 'disposition towards ...work was formerly affected by [a] sense of responsibility on the one hand and by ...fear of losing [a] job on the other. The security which the building operatives have enjoyed since the war has certainly tended to reduce the efforts of those among them who were formerly kept up to the mark by fear of unemployment.'

Such harsh pressures required substitutes under full employment, including incentive schemes, craft interchangeability, better coordination of labour as jobs closed and opened and the right spirit of cooperation:

> 'There is ...a long tradition of service ...and we believe that this will not be appealed to in vain. There is no general formula for improving morale except good leadership. It is a primary function of the representative organisations ...to provide such leadership.'

3.3.6   *Supply-side solutions*

Supply-side economics predominate, including training, commonality in architectural and management education, organisational innovations derived from scientific management, the availability of materials and components and technical innovations. These are left to employer and corporatist coordination regionally or nationally.

3.3.7   *Desirable future*

These again indicate aspirations for certainty, with abolition of variation clauses and the example of the USA. Variations clauses reflect post-war uncertainties of inflationary conditions and shortages which had undermined competition in building materials. Contractors had been sheltered

'from some of the main risks to which …business is normally subject in consequence of changes in prices of materials and wages.'

Failure to pursue incentive schemes, the increasing uncertainty and incompleteness of information in the client, the cost and inconvenience of variations and lack of standardisation of public contracts had exacerbated these effects. Cost variation clauses for materials and labour

'are no doubt essential, if undue margins for risk or gambling on price movements are to be avoided. It is clear, however, that as soon as conditions become more stable these clauses should disappear.'

The energy, vitality and attitude to work of USA construction are offered as a contrast. Complete preplanning by owners and architects, more efficient organisation of contractors and subcontractors, more extensive specialisation due to deskilling of crafts and more efficiency and skill of specialists are present. There is less rigid professionalisation, greater concern for clients, interchange of staffs of contractors and architects and well-educated site supervisors. Site facilities and services allow greater mechanisation. Standardisation is widespread. There are

'elements of uncertainty which are probably inconsistent with the full adoption of the American practice of preplanning; but we think there is – and in pre-war days was – a great deal of room for improvement here in this respect.'

3.3.8   *Comparison with Emmerson*

Emmerson's endorsement of construction's post-war performance has been noted previously. However, 14 years separate the inception of Phillips and publication of Emmerson, and hence questions arise as to the effect of Phillips' recommendations under the conditions of rapid change perceived by Emmerson. Certainly, clients, contracts and coordination, i.e. communications, are the focus of his report which shifts the emphasis to the 'construction industries' and makes 18 direct or indirect references to Phillips. Emmerson addresses efficiency as more active planning through the National Economic

Development Council (NEDC) develops. However, Emmerson operates within the evolving metropolitan corporatism of construction, which he seeks to strengthen by enhancement of the role of the national coordinating bodies:

> 'One result of my survey might well be a review of these arrangements, and an examination by the National Consultative Council of ways in which its usefulness could be extended and its authority enhanced.'

Emmerson addresses most of Phillips' key concerns. His comments reveal a more sophisticated appreciation of construction as befits a past Principal Secretary in the Ministry of Public Building and Works (MPBW) from 1944 to 1955 and previously Deputy Director General of Manpower in the Ministry of Labour and National Service and hence a 'Keynesian planner' of national significance. Construction's fragmentation existed at two levels, with weak coordination of the industry as a whole and fragmentation of production. He reflects the usual concerns: the weak attention of clients, the split of design and production and the general lack of cohesion:

> 'What is made clear is that if efficiency is to be increased, procedures and relationships must be improved ..'.

The 'three Cs', clients, contracts and coordination, reflect thematic continuity. Knowledge of the efficiency of the industry is weak: there is 'no statistical measure of efficiency in the construction industry as a whole'. Quarterly statements of the value of work are

> 'a useful guide to trends in the volume of work done but it would be wrong to regard them as an index of efficiency.'

Phillips' recommendations on incentive pay were unrealised. A total of 14% of operatives within the construction industry were on incentive schemes compared with 42% of the workforce in manufacturing, although other supplementary payments had proliferated owing to labour shortages. Emmerson notes a requirement to extend the use of scientific management and costing methods. Abolition of variations (firm price tendering) had arrived in April 1957 subject to conditions:

> 'one that the work was thoroughly planned in advance, and the other that the estimated contract period should be not more than two years.'

This had been successful where the conditions had applied.

Professional and management education, especially joint education, and training generally are still preoccupations. A special conference at RIBA in early 1956 resolved that[18] 'the industry could improve its standards and raise its productivity by interrelating the training of its constituent administrative branches.'

A scheme had emerged from this conference for a

'staff college of advanced building technology in London intended to serve in the first instance the contracting side …This scheme has hung fire …'.

Emmerson refers to the Hall inquiry[19] into the establishment of a scheme of joint education for construction professionals under the auspices of the National Joint Consultative Committee (NJCC), a 1956 British Institute of Management (BIM) joint study and conference with the Board of Building Education which had been established by the NFBTE under the chairmanship of Sir Hugh Beaver, Emmerson's predecessor at the Ministry of Public Buildings and Works, and a London and Home Counties Regional Advisory Group study and report from 1957. He does not refer to 1959 or 1962 Institute of Building (IOB) initiatives which led to the Building Management Notebook[20], the first distance learning course for construction managers, NJCC sponsored attempts to develop the theme of communication in 1958[6] or a study of construction management roles by the Board of Building Education at the IOB[6].

He notes that Government interest since the recommendations of the 1956 Building Apprenticeship and Training Council report had been fitful

'and there has been no-one charged with keeping under review the arrangements for training at different levels and their progressive development.'

He recommends action again by the National Consultative Council of the Minister of Works.

The importance of the public client and the public interest in greater efficiency remained, and Emmerson details a range of initiatives for coordinating public procurement across a range of departments, predictably discussing the Consortium of Local Authorities Special Programme (CLASP) but also mentioning work on the coordination of sizes of building products which could yield significant rationalisations and efficiencies, attempts in the context of NEDC planning to coordinate Government department procurement more effectively and the continuing importance of standardisation and prefabrication.

3.4   Discussion

At the start it was argued that construction's problems and solutions are recycled in different forms from report to report. The recycled solutions take on a more complex and apparently encompassing form[21], business process re-engineering (BPR) 'perfecting' the scientific management offered during the

1950s but which according to Emmerson failed to take off. This failure is self-evident from Phillips: construction had serious enough problems recon-structing its labour force after the dislocation of World War II, coping with labour shortages, new expectations over 'rebuilding Britain' and the tech-nical and other innovations detailed by Emmerson.

When political action moved towards a more active concept of planning in 1961 the methodology of establishment networking was exposed as inade-quate to the scale of investment and speed of innovation required[4]. Emmerson was appointed in 1961, and reported quickly in 1962. However, the perceived scope of the problems created by construction for political ambitions led to the replacement of The Minister for Public Building and Works, Lord Hope, by a more energetic and ambitious individual, Geoffrey Rippon, before Emmerson was published. He instituted a wholesale restructuring and sophisticated managerialisation of the MPBW[22, 23], probably beyond the level of construction itself but reflecting in its approach a conviction of the relevance of systematic management and the use of public procurement to drive innovation and productivity. Hence, there was a political outpacing of establishment networking and cajoling towards managerialisation and the coordination of education and production through the NJCC[22].

3.5   Conclusions

Emmerson reveals marked continuities from Phillips, but discontinuities can be shown in terms of the non-realisation of many of Phillips' objectives and the exacerbation of problems. Typically, further reports were recommended. Contracts were dealt with by Banwell[24], a committee appointed by the Minister, and coordination by the Tavistock researchers of the Building Industry Communications Research Project (BICRP) under the heading of 'Communications' and the auspices of the NJCC. This split contracts and their problems from the analysis of interorganisational relationships and project dynamics, with 'communications' treated non-prescriptively for the first time. Research became the focus of a separate report[25], again ministeri-ally driven, which destabilised the BICRP[14]. This led to the establishment of the Building Research Establishment (BRE) from the existing Building Research Station and BIRA, later CIRIA, from the Civil Engineering Research Association, reflecting continuity of Phillips' concerns about dissemination which still created difficulties.

The overall weakness of Phillips flows from the origins of the Committee in the world of wartime collaboration and planning reflected in Simon[3] which had assumed the attainment of sufficient stability around the project to make the contract work. Consequently, there is an aspiration to coherence in a national building system which, given labour market changes, growth of client uncertainty and technical change, was probably unrealisable in 1950. The false coherence of the report results from the segmentation of problems and the juxtaposition and elision of solutions. The uncertainties revealed by Phillips are replicated in Emmerson who summarises the extent of their proliferation in spite of attempts to improve coordination and management, revealing where Phillips' putative solutions fell short. Continuities exist in the attempt to evolve the basis for corporatism, the reliance upon a rationalistic and limited managerialism and invocations to collaboration and better coordination. In this sense Phillips constitutes a template for most of the postwar reports through which Government has sought to influence construction.

3.6   References

1   *Phillips Report on Building* (1950) Working Party Report to the Minister of Works. HMSO, London.
2   Pettigrew, A. M. (1985) Contextualist research: a natural way to link theory and Practice. In: *Doing Research that is Useful for Theory and Practice* (ed. E. E Lawler III). Jossey-Bass, San Francisco.
3   Simon, Sir Ernest (1944) *The Placing and Management of Contracts.* HMSO, London.
4   NEDO (1964) *The Construction Industries.* HMSO, London.
5   Wild, A. (2001) The Phillips report of 1950. In: *Proceedings of the 17th Annual Conference of the Association of Researchers in Construction Management*, University of Salford, 7–9 September.
6   Higgin, G. & Jessop, N. K. (1965) *Communications in the Building Industry.* Tavistock Publications, London.
7   Andrews, J. & Darbyshire, Sir Andrew (1993) *Crossing Boundaries.* CIC, London
8   Boyd, D. & Wild, A. (1999) Construction projects as organisational development. In: *Proceedings of the 15th Annual Conference of the Association of Researchers in Construction Management (ARCOM)* (ed. W. Hughes), Liverpool John Moores University, 15–17 September.
9   Schon, D. A. (1971) *Beyond the Stable State.* Temple Smith, London.
10 Reed, M. (1989) *The Sociology of Management.* Harvester Wheatsheaf, Hemel Hempstead.
11 Latham, Sir Michael (1993) *Trust and Money.* HMSO, London.
12 Latham, Sir Michael (1994) *Constructing the Team.* HMSO, London.
13 Taylor, A. J. P. (1971) *English History 1914–1945.* Pelican Books, London.
14 Crichton, C. A. (1966) *Interdependence and Uncertainty: A Study of the Building Industry.* Tavistock Publications, London.
15 Simon, Sir Ernest (1948) *The Distribution of Building Materials and Components.* HMSO, London.
16 Simon, Sir Ernest (1945) *Rebuilding Britain: A Twenty Year Plan.* New Left Books, Victor Gollancz, London.
17 Powell, C. (1996) *The British Building Industry since 1800. An Economic History.* E & F Spon, London.
18 Emmerson, Sir Harold (1962) *Survey of Problems Before the Construction Industries.* Report prepared for the Minister of Works. HMSO, London.

19   Hall, Sir Noel (1964) *Joint Committee on Training in the Building Industry*. NJCC, London.
20   IOB (1962) *Building Management Notebook*. Institute of Building.
21   Green, S. (1998) The technocratic totalitarianism of construction process improvement. *Engineering, Construction and Architectural Management*, 5(4) 376–86.
22   Hislop, M. (1971) The industry and the market; the industry and the professions; development of the industry. In: *Construction Management in Principle and Practice* (ed. E. F. L. Brech). Longman, London.
23   HMSO (1963) *The Reorganisation of the Ministry of Public Buildings and Works*. Command 223, December. HMSO, London.
24   Banwell, Sir Harold (1964) *The Placing and Management of Contracts for Building and Civil Engineering Work*. HMSO, London.
25   Woodbine Parish, D. E. (1964) *Building Research and Information Services*. HMSO, London.

Chapter 4

Survey of Problems Before the Construction Industry:
A Report Prepared by
Sir Harold Emmerson (1962)

Krisen Moodley & Christopher Preece

4.1 Report brief

In the autumn of 1961, Lord Hope, the Minister of Works, asked Sir Harold Emmerson to undertake a quick survey of the construction industry. This survey was to focus on the problems that were most significant for increased efficiency within the industry. There were no other formal terms of reference.

Lord Hope did, however, write to a variety of organisations within construction, setting out the broader context of Emmerson's work. He indicated that the Chancellor had put into place economic policies that sought to restrain the increases in demands on the construction industries. This was in the context of increasing demand over the next 10 years. Lord Hope sought to identify measures that both the Government and the industries could undertake to meet the expected increase in demand with the greatest efficiency. In instructing Emmerson to undertake an independent review of the problems facing the industry, Lord Hope wanted to identify those issues that were of greatest importance to productivity and how to tackle them. This created the context for the survey into problems within the construction industry.

The results of this survey formed the basis of what is described as the Emmerson report into the problems of the construction industry.

4.1.1 Economic and political conditions prior to the Emmerson report

The Emmerson report was the first major report into the construction industry to be published since the Simon report in 1944. The changes that had taken place in the United Kingdom since the end of World War II were the drivers for the need for a report on the construction industry. The country had gone through a period of reconstruction and seen the introduction of the welfare state. The economy had to all intents returned to normal by 1951, although there were issues such as inflation, nationalisation and currency value that still had to be addressed.

The main aims of policy under the Conservatives during the 1950s were to ensure full employment, stable prices, balanced development in different parts of the country and provision for social security on the Beveridge model. There was greater emphasis on market freedom, a high priority was given to price stability, there was a commitment to full employment, more use was made of monetary weapons and there was a general dislike of anything that suggested centralised planning[1]. These policies set the economic agenda for the period up to the start of the Emmerson report.

The decade that preceded Emmerson was not as dramatic as post-war reconstruction or the 1960s in terms of economic changes. The outbreak of the Korean war saw a shift in economic emphasis towards rearmament and a mini-boom. This was immediately followed by a recession in 1952. A mini house-building boom in 1953 was part of the strategy to get the economy moving again. The economy improved by 1955 but with the associated fears of inflationary pressures. Attempts were made to introduce a credit squeeze, increase taxation and reach agreements with industry to hold down prices and wages. The Suez affair took place in the autumn of 1956 and led to foreign exchange difficulties. The impact of Suez was relatively minor on the domestic environment, and the economy continued to expand at a slow pace. In 1957, Chancellor Thorneycroft tried to fight inflation by tackling public spending, bank credit and the stock of money, in spite of the implications to employment. In 1958, unemployment rose to above 2% with the demand for reflationary measures. Thorneycroft resigned as Chancellor as the anti-inflationary measures he wanted to introduce were unpopular. By the end of the 1950s the economy had returned to the overheated condition that had started the decade. The 1950s are best remembered as a decade of low unemployment and comparative stability of prices. Table 4.1 shows some of these economic trends.

At the start of the 1960s a difficult situation was emerging in the economy. Growth at the end of 1959 had been faster than expected. The Government, although reluctant to introduce deflationary measures, increased interest rates, and hire purchase restrictions were reimposed. By the end of the year, growth had declined and the Government started measures to reduce the impact of interest rates. In 1961 the expected pressure on demand for the following year alarmed the Government. Forecasts suggested that there was a need to reduce growth. In the Finance Bill of July 1961, interest rates were raised 2%, Government expenditure was limited and foreign investments were scrutinised. There was also a call for a pause in wage increases, and one was imposed on the private sector. The National Economic Development Council (NEDC) was created in July 1961. The NEDC was charged with planning long-term growth. The NEDC would look at issues such as growth, productivity, income policy, etc. The Emmerson survey comes as no surprise because it deals with efficiency and long-term planning. Its themes fit with the economic issues prevalent at the time.

Table 4.1 Changes in the allocation of resources, 1955–61 (increase from year to year in £ billion at constant 1985 prices) according to Economic Trends National Supplement (1990 edition).

	1955	1956	1957	1958	1959	1960	1961
Consumers' expenditure	4.4	1.0	2.3	2.7	4.9	4.6	2.7
Public authorities' final consumption	–1.2	–0.5	–0.8	–1.2	0.8	0.9	1.6
Gross domestic fixed capital formation	1.3	1.1	1.4	0.2	2.0	2.6	3.0
Value of physical increase in stocks and work in progress	1.5	–0.4	—	–0.8	0.6	2.5	–1.9
Exports of goods and services	1.8	1.3	0.8	–0.5	0.9	1.9	1.1
Total final expenditure	8.9	3.0	4.3	0.8	9.5	13.0	6.3
Imports of goods and services	2.8	0.1	0.8	0.3	2.1	4.0	0.7
Gross domestic product (at market prices)	5.9	2.9	3.5	0.5	7.4	8.9	6.7
Gross domestic product (average at factor cost)	5.4	2.1	2.6	–0.3	6.4	9.2	4.7

4.1.2   Construction industry prior to 1961

In the immediate aftermath of World War II there was an obvious need for reconstruction. These activities kept the industry from the scrutiny it required. By 1950 the country was returning to normal in economic terms and construction was now required to operate on a normal footing.

The development of the welfare state had seen an expansion in the demand for buildings for the creation of these new social services. In addition, social housing provision was also increased. In working towards a full employment policy, the demand for construction followed the characteristics of a growing economy rather than a stagnating economy. The demand for construction was based on economic and social need for what was seen as a period of expansion and redevelopment.

The development of the welfare state also increased the role of the Government as a major construction client. There was an increased level of investment in buildings by public authorities. Local authorities in particular dominated the housing market and schools sector. The greater involvement of the public sector in construction also meant that the industry was more susceptible to changes in Government economic policy. In 1961 the Government had initiated a move towards a planned economy with the NEDC, and its influence would be extended further.

The Governments over the period prior to the report saw the building industry as an ideal tool to control the general level of economic activity. This created particular difficulties in ensuring the most economic use of construction resources (an issue addressed in the Emmerson report). Efficient and economic use of resources required a steady and balanced flow of demand. The increased influence of the Government-controlled sector brought with it an expansion of short-term planned projects. This is a characteristic of public control. Innovation in contract development was also hampered by public sector influence. The most dominant force in the period up to the report was the Government.

It was recognised that there were unsatisfactory relationships between the construction professions and the rest of the industry during the inter-war years. This situation had been highlighted in the Simon report but still persisted within the industry in the 1950s and early 1960s. Bowley[2] indicated that it was only as late as 1955 that the representatives of the professions and the industry began to admit that all was not well. Doubts were expressed over the suitability of the organisation of professions and an industry that grew up in the nineteenth century being suitable for the changing social and economic conditions.

The nature of construction contracts had altered little since the inter-war years and was dominated by open competitive tendering. The Ministry of Works had used selective tendering since 1939 but remained an isolated example. The Simon report had suggested that open tendering be abolished in the strongest possible terms. As late as 1957 the Ministry of Housing and Local Government had actually discouraged any modification of traditional contract procedures[2]. New Model Standing Orders were issued to local authorities in England and Wales to confirm that open competitive tendering was to be enforced. The strength of local authorities in upholding what they saw as a tried and tested system blocked any effective change in contract procedure. Layton[3], in her report on building by local authorities, also identified problems with open competitive tendering but indicated that the authorities were restricted by the need for accountability. She also indicated that competitive tendering could be improved if local authorities took greater cognisance of the needs of builders. Layton's report suggested that local authorities could improve contract practice by being more efficient.

In 1957 the Government attempted to reintroduce the principle of fixed price contracts. This was part of broader economic measures brought about by a general credit squeeze. Fixed price contracts led to widespread criticism of the system in use. It was argued that fixed price contracts were impractical unless projects were fully preplanned, permissions and permits were granted and completely detailed drawings for accurate estimating were provided. There was a perceived inefficiency on the design side, particularly with the failure to deliver drawings, avoid variations and so on, which was seen as an obstacle to public policy. The Royal Institute of British Architects (RIBA) initiated a study into the procedure and efficiency of architects' offices in recognition of the serious nature of the problem. Emmerson in his report devotes time to the issue of architects and the problems that arise from fixed price contracts. Fixed price contracts were also difficult to implement because of the changing economic conditions.

The nature of construction projects, unlike many aspects of the industry, underwent change during the 1950s and early 1960s. The increasing development of large-scale organisations tended to lead to larger and more complex construction projects. This posed a risk to firms as the projects often entailed a significant level of financial risk. Specialist engineers began to play a more prominent role in more complex facilities. It became more common for engineers to own or work for contractors in the building industry. There was also an increase in the need for more pleasant and appropriate conditions for living and working. This reinforced the role of the architect as the lead consultant on buildings. The architects, as indicated earlier, faced a great deal of scrutiny as to whether they were capable of continuing as lead consultants as the demands on projects changed. Emmerson focused on the architects in his report as he saw them as crucial to the way in which construction was implemented. As a body, the RIBA was extremely powerful and many of its members were employed in local authorities in positions of influence.

4.1.3   The scope and purpose of the enquiry

In the autumn of 1961, Sir Harold Emmerson was asked to undertake a quick survey of the construction industry. He had no formal terms of reference, as indicated earlier, but was supported by the Ministry of Works. It was expected that he would report by the end of 1961. The organisations consulted were Government departments, professional associations, utility boards and industrial organisations. The report excluded direct labour schemes of local authorities or overseas programmes.

Emmerson proceeded on an informal basis, partly because this seemed to be the best way to elicit information and partly because of the time constraints. During the survey, Emmerson stressed the independence of his role in reporting to the Government. He enjoyed the cooperation of all those who participated but regretted that the time constraints only allowed him to see a relatively small sample of people. It is also apparent that as a quick survey it was difficult to explore all sections of the construction process. Construction had over 50 materials industries, a complex system of distribution of supplies, numerous professions and a vast number of individual firms, from one-man firms to multinational enterprises. Emmerson indicated that his report was based on the organisations that represent the interests of the various parts of the industry rather than the complete process and individual firms.

As the title of the report suggests, it was intended to highlight problems within the construction industry. In his introductory statements, Emmerson does, however, indicate that the industry had made a strong post-war recovery. The construction industry had shown itself to be flexible in meeting demand, had introduced new materials, had increased mechanisation and new construction techniques, had shown a steady rise in output and had avoided major industrial disputes. He also indicated that there were obstacles to progress and many of the difficulties the industry was facing were not of its own making. Construction was, however, more in the public eye, and their deficiencies were more readily seen. On the whole, construction did not make enough of its own achievements.

In general, the purpose of the Emmerson survey was to look at problems related to efficiency within the industry. At this time there was no easily identifiable measure of efficiency, and the report was constructed largely on a qualitative basis along broad lines.

4.2   The report

4.2.1   *Confidence and continuity*

Emmerson identified that the construction industry had been used by successive Governments as a convenient instrument for regulating the economy and would continue to do so. The construction industry did not create the demand for its services and was governed by the general level of economic activity and by Government policy in relation to social services and capital investment. No industry could function efficiently with successive periods of overloading and underemployment.

The construction industry indicated that, for greater efficiency to exist, the Government would have to adopt as a main feature in its policies a steady and expanding construction programme for some years ahead, to keep pace with a steady rate of growth in the economy as a whole. This would help remove the 'boom and bust' cycle that influenced the industry. The professions indicated that they were also responsive to a continuity of demand, while some sectors in the materials industries preferred full capacity. The construction industries felt the need for greater continuity and the importance of a long-term outlook for greater efficiency.

Emmerson identified that a long-term approach was essential for the confidence of the industries. Where a lack of confidence existed, the firms were unlikely to commit to investment in new plant and equipment, and innovative construction techniques were unlikely to emerge. Larger firms were more likely to have the resources and initiatives to undertake risk. The medium and smaller firms, which make up a large part of the industry, were unlikely to take a risk unless there was confidence in the long term. Confidence also influenced recruitment and the potential to get quality recruits.

New Government initiatives would hopefully encourage continuity and confidence. The Government had recommended that local authorities plan ahead.

4.2.2   *Relations between the building owner, the professions and the contractor*

Emmerson indicated that efficiency in building operations was dependent on a clearly understood division of responsibility between the various parties, adequate numbers of professionals, the quality of the professions and the quality of their relationships. In the case of private sector clients, Emmerson generalised that they were largely dependent on the architect for advice and guidance. In the private sector it became apparent that the architect had a major influence on the quality and standard of work. Nearly 20 000 architects practised in the country, of which half were employed in some 3000 private practices. About three-quarters were employed in practices with less than six staff. More public sector work would be put out to private architectural practices as the long-term building programme took shape. Emmerson suggested that, because many of the architectural firms were small, they did not meet modern requirements in matters such as organising ability, cost control and office management. He also indicated that it would better if there were a consolidation of the smaller practices into larger units.

The RIBA was aware of the need for change and was actively engaged in studying ways of increasing the efficiency of the private architect. Emmerson suggested that this could not be left to education alone, but something had to be done to improve organisations and methods. The RIBA had launched its own survey of the organisation, staffing and costs of the architect's office. The survey[4] was underway during the writing of the report. The RIBA was considering matters such as management guides, standard forms for procedures, uniform methods of cost control, etc. In general the RIBA recognised that it must act as a coordinating body in the interests of efficiency for architects and the public.

The survey suggested that there was a great deal of criticism of the lack of cohesion between the architect, other professionals and the builder. In the civil engineering sector there were better relations between the civil engineer and the contractor. In building it appeared that there was a greater lack of confidence between the architect and the builder. Emmerson identified that in no other significant industry was the responsibility for design so far removed from the responsibility for production. Methods of training and the codes of conduct aggravated this lack of cohesion for members of the professional bodies.

Another factor that influences relations is the restrictions imposed on architects and surveyors in the freedom of movement if they are not to lose their professional status. While restrictions safeguard the independent status of architects and chartered surveyors, they do inhibit closer relations between the parties. These advantages are self-evident within civil engineering. Emmerson suggests that some alternative way be sought to bring builders and their professional associates closer.

Emmerson praises the work done by the National Joint Consultative Committee (NJCC) of the RIBA, the Royal Institution of Chartered Surveyors (RICS) and the National Federation of Building Trade Employers (NFBTE), since its appointment in 1955, in the examination of problems affecting architects, surveyors and builders. He also suggests that the Committee should usefully consider how to secure better coordination and cohesion between the architect, surveyors and various other consultants brought together at the design stage, and between the professions and contractors. A separate committee to look at this issue should be instituted.

4.2.3   The placing and management of contracts

This report was the first to review contract arrangements in the construction industries since the Simon Committee report in 1944. The recommendations of the Simon Committee were adopted as good practice in the 1950 Working Party and the Anglo-American Productivity Team. In 1952 the Ministry of Works invited the RIBA to take the lead in reducing costs and reviewing contract arrangements. A joint committee with the RIBA, RICS and NFBTE was formed for the enquiry. The Committee, under the chairmanship of Howard Robertson, confined its consideration to tendering procedure and reported in 1954 on the subject. In general the Committee agreed with the Simon report and recommended, *inter alia*, the adoption of selective tendering and a return to fixed price contracts. The Robertson Committee did not, however, have representatives from local authorities and was not able to obtain the cooperation of all local authority associations in their work.

The Robertson Committee's recommendations on selective tendering were put in place through a code of practice for selective tendering issued by the NJCC in 1959. The Ministry of Works returned to firm price tendering for Government work in 1957. This was subject to two conditions: one that work was planned in advance, and the other that the estimated contract period was for not more than 2 years. Emmerson found that firm price contracts required a high degree of efficiency and worked well under specific conditions. He also recognised that, with regard to fixed price contracts, no building owners in the public sector adopted the lead given by the Ministry of Works.

Emmerson found that, in spite of the reports produced in the 1950s, there was still a failure to adopt procedures that were designed to secure for the building owners the benefit of effective competition between firms of similar standing. Open tendering was still common and this was working against firms that maintained high quality, and the owner was not necessarily gaining the best value from the lowest-value tender.

Emmerson found that there had been other developments in the management of contracts. Simon and Robertson had recommended that the main contractor should be in full control of all subcontracts and should be enabled to plan the whole of the works. It was recognised that some highly specialised processes had to be conducted by nominated subcontractors. The Government had also adopted an approach that set out to reduce prime cost items on buildings, leaving the main contractors to select their own specialists. This was in contrast to the private sector where there was a growing trend towards a higher proportion of work being let out to nominated subcontractors.

Emmerson identified the growth of the 'package deal' where the building firm provides a complete building directly for the client. The contractor provided all the professional and construction services required for the building. The contractor was responsible for design and could therefore use techniques that took advantage of the contractor's expertise and plant. It was claimed that this system enabled better planning and execution of building, and ultimately lower costs to the building owner.

Emmerson also found that other forms of placing and managing contracts were being introduced, such as serial contracts. Tenders were being invited for one job with the intention of engaging the selected contractor for successive jobs in a continuing programme. These methods required earlier collaboration between the contractor and design teams. Emmerson did, however, indicate that public bodies had to be in a position to withstand criticism in the selection of contractors when they departed from standard procedures, although he did point out that some of the principles laid out in the two earlier decades were no longer valid.

Emmerson argued that there was a need for a far-ranging enquiry into the management and placing of contracts. He suggested that a study by an independent chairman be undertaken to look at contracts in the context of modern conditions. (This led to the Banwell report.) He suggested that it should include the professions, the building industry and local authorities. He indicated that it should follow the same format as the Simon Committee in 1944 and bring the whole of construction under review. He suggested that the following should be reviewed:

- the operation of the retention money system;
- experiences of the fixed pricing system;
- the possibility of introducing a common form of contract for both building and civil engineering work;
- how the conclusions/principles of the inquiry could be translated into practice, at least in the public sector.

4.2.4   Apprenticeships and training

Emmerson was 'assured' that the civil engineering contracting industry was 'increasingly alive to the importance' of training but recognised that the nature of the work often made it difficult for continuity of employment and that the lack of variety did not make it desirable for craft apprenticeships.

However, Emmerson refers to the National Apprenticeship Scheme (1945) and the Building Apprenticeship and Training Council (1956), asserting that a number of issues still required addressing in 1962. These included, for example, the length of various craft apprenticeships, new opportunities in schools and technical colleges, the possibility of transferring apprenticeships between firms to give greater variety and the costs of training.

Emmerson observed that building techniques had changed and that new skills would have to be developed. New definitions of crafts and skills would be required to take into account modern technological and operational developments.

The report emphasised the need for greater understanding of employment conditions and the reasons for so many non-indentured learners. Also noted was the need to examine the degrees of skill required given the progress made with new materials, prefabrication and industrialised building methods. Emmerson saw a role for Government departments in helping the National Joint Council for the Building Industry, but provided little in specific recommendations. The report suggested that rationalisation of apprenticeships would help avoid demarcation disputes and in reorganisation of trade unions in the industry. While Emmerson recognised that construction had been receptive to new techniques and materials, he concluded that disputes had arisen owing to changes in 'division of responsibility' between different crafts.

4.2.5   Conditions of employment

Emmerson reported that there had been a 'noticeable absence of major industrial disputes' and that any labour problems had been resolved without Government intervention. This was seen as positive for improved efficiency and productivity.

The National Joint Council for the Building Industry and the Civil Engineering Conciliation Board had implemented a number of recommendations on holiday pay, payment for time lost owing to bad weather and a 'guaranteed minimum weekly pay packet'. This had led to a 'substantial' reduction in casual employment.

Emmerson asserted that a great deal more could be done to improve conditions. The larger firms had increased their permanent labour force but were still suffering from turnover of labour which could vary from district to district and from firm to firm. Emmerson asserted the need for a high proportion of operatives to be employed on a regular and continuous basis. This was seen as essential to improve public perceptions. The idea that employment in construction was 'casual' seriously affected recruitment 'from apprentice up to management'. Emmerson suggested that a study should look into the effect of continuity of employment on the various forms of contracting. The findings would enable the formulation of a policy for increased stability of employment.

The report also considered working conditions and concluded that it was more difficult to provide welfare facilities comparable with those of factory employment, but the point of 'attack' should be those areas where there was 'greatest contrast with factory employment'. No practical recommendations were given as to how this would be done.

Emmerson reported that no agreement had been reached on incentive payments which had been seen as essential for increased output by the 1950 Working Party. He suggested that more information was needed about 'actual conditions in the industry, the effect on production of incentive payments' and the 'desirability of higher rewards for skill'.

4.2.6   A need for competent management

The Emmerson report 'reinforces' the findings of earlier work of the Board of Building Education, which had been set up by the Institute of Building and the NFBTE, and their 1956 joint report with the British Institute of Management by Sir Hugh Beaver. Emmerson suggested that a number of the recommendations of this work had not been fulfilled, in particular in the area of site management. He specifically highlighted the growing need for management training to be spread more widely, particularly for the small and medium-sized operators. The advantages in adopting modern methods of 'cost control, planning, programming and site management' were emphasised.

4.2.7   Further action required on training and higher education

Emmerson believed that the leaders of the construction industries were 'aware of the need for reform and were taking action'. However, no body had been charged with reviewing arrangements for training at different levels. While the Ministry of Labour was in touch with a new joint committee on apprenticeship and the Ministry of Education was interested in proposals on new forms of training, there was 'no clear line of responsibility'.

Emmerson recommended that, in matters concerning management training and higher education for supervisory, administrative and technical staff, the National Consultative Council of the Minister of Works should consider further action by Government in promoting greater interest in the courses already available.

4.2.8   A need for more research

Emmerson discovered that there was increased interest in research, particularly in civil engineering, with the Institution of Civil Engineers and the Federation of Civil Engineering Contractors setting up bodies of their own.

This was to improve cooperation between the industry and the Building Research Station (under the Department of Scientific and Industrial Research). He found, however, that most of the research going on in the industry was primarily in the large contracting organisations that could afford research laboratories. This research was

> 'directed to materials, the design and performance of structures, engineering services and environmental conditions such as sound insulation and lighting.'

Emmerson commented that more interest could be generated from the construction industry and professions if research were to deal with site organisation, work study, cost control, conditions of employment and human relations. He asserted that this would require more money and staff but that 'there can be no doubt as to the value to be gained'. He emphasised that there was no lack of urgent work needed in examining methods and organisation at different levels.

4.2.9   *Public interest in increasing efficiency in construction*

Emmerson's report clearly demonstrates the level of concern of central government in the need to improve the efficiency of these industries, as so much of the public finances was being spent on building and civil engineering. The new National Economic Development Council would help a programme of expansion, and the concern was whether construction could meet the increased demands.

Emmerson clearly saw the role of Government in influencing the changes necessary, not in terms of imposing controls but by creating new relationships between Government departments and the industries in trying to engage construction in increasing efficiency itself. The solutions to this, according to Emmerson, lay in the experience within the works directorates of the Government departments. Professional heads and chief architects of departments responsible for schools, hospitals and housing could meet regularly to exchange experience on procedures, drawing office methods, etc. This was seen as possibly the way forward for local authorities and other public bodies.

Emmerson recommended that 'strong guidance' be given to local authorities and other public bodies in carrying out their construction programmes. Guidance had been given to the universities by the University Grants Committee. This should be extended to programmes in the social services, i.e. hospitals.

Emmerson asserted the need for greater standardisation of materials for building. He suggested that a central driving force was needed to get things done, and that it should not be left to the manufacturers. He also argued that a choice should be made between the systems of modular coordination, i.e. standard units of measurement for building components. Greater urgency was felt with the possibility of entry into the Common Market.

The report returned to the issue of dissemination of research in the industry and the importance of ensuring implementation of better materials and methods. The question as to who would see that this was carried out was posed. The new Council for the Construction Industries was encouraged to develop a 'closer relationship with central government' to improve communication on policy and programmes.

4.2.10   *The report's key recommendations*

- More standardisation of materials and components.
- Efficiency in operations depends on the quality of relationships and better coordination between building owner, the professions (architect, surveyor, engineer) and the contractors and subcontractors.
- A review was needed of the placing and management of contracts by Government committees with the cooperation of local authorities.
- A working party should examine the different systems and practices in Scotland and look at ways of unifying these.
- Much would rely on the numbers and quality of personnel at all levels. Government's role would be to give practical help and encouragement in the use of training facilities.
- More research was needed into the economics of building. Organisations concerned should be asked to advise on priorities.
- Government should support the establishment of a technical information service for architects, surveyors and builders.
- A new form of relationship with Government was needed and existing arrangements for consultation should be reviewed.

4.3   The impact of Emmerson

The Emmerson report was undertaken at an interesting point in the economic and social development of the UK. The country had gone through a period of post-war reconstruction and the creation of the welfare state. By 1960 the pressures of inflation and exchange problems had created a more turbulent environment to manage. The policy towards the creation of full employment and a social welfare system had increased the public expectation. Western Europe had largely recovered from World War II, and its economy was powering ahead, increasing the competition for the UK. The colonial influence of the UK was also declining, and it faced challenges in the majority of its overseas markets. The UK economy in the 5 years prior to the Emmerson report had fallen behind its key competitors; it had had steady growth but faced challenges in the new decade.

The UK construction industry had been largely isolated from criticism during the preceding years of the report. The last major report into the construction industry had been the Simon report. The Simon report raised many issues, but much of its impact had been diluted by the needs during the war and the post-reconstruction period. It was only from 1955 onwards that there was a growing recognition that all was not well with the construction industry. In 1957 the Government attempted to introduce fixed price contracts, and it was at this point that matters seemed to come to a head. Construction was an integral part of Government economic policy, but the industry was not in a position to deliver what was needed. The Emmerson report was needed to highlight some of the problems within the industry.

The key issue within the Emmerson report is the management and placing of contracts. Contract procedure had not altered since the 1930s and the Simon recommendations had not been implemented. The 'system' had not changed in any serious way, in spite of the problems that existed within it. The system of placing contracts and the relationships around them had been designed in the nineteenth century and were not suitable for the twentieth century.

At the root of the problem was the manner of placing contracts. Open competitive tendering was the norm, but this system was inefficient and did not meet the needs of industry. Problems in the management of cost, variations and design also emerged from this system. Emmerson found that Robertson's recommendations on selective tendering and fixed price contracts had not been adopted. It also became apparent that the Ministry of Works and the local authorities were working at odds with each other, with the latter pursuing a policy of open tendering. While Emmerson did not have time to include the opinions of local authorities in his report, it is clearly implied within the report that they needed to be addressed as key influences within construction.

Emmerson suggested that an independent inquiry specifically into the management and placing of contracts be considered. He highlighted areas of concern such as the retention system, fixed pricing and common forms of contract. This inquiry was to be far reaching and cover every sphere of construction. In terms of contribution, this is the biggest influence of the report as it laid the foundations for the Banwell report. Emmerson's contribution was that he highlighted that all was not well within the construction environment in terms of management and placing of contracts.

Directly related to the placing of contracts was the issue of relationships within the construction industry. Emmerson found that relationships were strained, and construction was one of the few industries where there was such a separation between design and production. It became obvious that the architect was crucial to the entire construction process. Emmerson found that architects were the major influence on standards and quality in the private sector. Layton[3] found a similar influence within local authorities. However, because many private architectural practices were small, Emmerson questioned whether they were capable of meeting the requirements of a modern industry. The key areas that were lacking included organisational ability, cost control and office management. He suggested consolidation of architectural practices to improve the situation. At this time the RIBA was also in the process of studying ways of improving efficiency as a great deal of criticism had been levelled at architects since the reintroduction of fixed price contracts. The RIBA study[4] influenced the emphasis that Emmerson placed on professions in his report.

The second part of the examination of relationships related to how all the parties operated. Emmerson found that there was a lack of cohesion between the architect, the other professions and the builder. Relations were particularly poor between the architect and the builder, unlike civil engineering. He found that methods of training and codes of practice created further problems. There was a lack of understanding of the roles each had to fulfil. Freedom of movement with regard to professions also prevented any cross-fertilisation of ideas. Emmerson advocated another committee to look at how to secure better coordination between the professions and with contractors. The report highlighted that the interaction between the parties in construction needed review.

4.4   Conclusions

The Emmerson report into the problems of the construction industry was not groundbreaking. It is probably best to start with the key problems with the report. Firstly, the report was to be delivered in a relatively short time frame. This did not allow it to have the depth that was required for an industry review. Secondly, the report did not consult extensively and in particular missed out on the participation of local authorities, which were a key influence on the construction industry at the time. Local authorities were at the time the key party in perpetuating the traditional system.

What, then, was the contribution of Emmerson? The report identified that Government interference in the industry was not good for efficiency. It also became evident that many of the Simon report recommendations were still not in place. The nature of the industry had not changed dramatically, although the influence of the private sector was increasing. Recruitment and education could be improved in the industry.

The major influence of Emmerson was that he identified that construction was operating a nineteenth century system in the twentieth century. The management and placing of contracts were archaic and innovation was stifled. The professionals and contractors were operating in an atmosphere of distrust that did not foster good relations or contribute to improved working practices. Emmerson realised that there was an urgent need for an overhaul of the system in place and recommended a wide-ranging review of the management and placing of contracts. In 1964 the Banwell report followed with these issues as a central theme.

The Emmerson report did not directly affect many of the parties within the construction process. Its contribution was fairly limited, although it did act as the catalyst for the more significant Banwell report.

4.5 References

1 Cairncross, A. (1995) *The British Economy since 1945*, 2nd edn. Blackwell, Oxford.
2 Bowley, M. (1966) *The British Building Industry*. Cambridge University Press, Cambridge.
3 Layton, E. (1961) *Building by Local Authorities*. Allen and Unwin.
4 RIBA (1962) *The Architect and his Office*. Royal Institute of British Architects.

Chapter 5

The Placing and Management of Contracts for Building and Civil Engineering Work: The Banwell Report (1964)

Cliff Hardcastle, Peter Kennedy & John Tookey

5.1 Introduction: the UK pre-Banwell economic and social environment

In June 1947, General George Marshall, the United States Secretary of State, outlined his famous 'European Recovery Program' (ERP), subsequently known as the Marshall Plan. That infusion of aid into Europe sparked a dramatic expansion in economic activity around the world in the 1950s, particularly in the United States and Europe. Rebuilding the destruction wrought by World War II was of primary concern; in Europe the ERP subsidised infrastructure renewal in both the former allies and the defeated axis powers. Expanding markets around the world boosted demand for British products, manufacturing revenues and confidence in the population as a whole.

After a decade of continuous expansion, in the general election of 1959 Harold Macmillan famously told the British electorate, 'You've never had it so good'. Indeed, full employment was a reality in the country at the time as successive Governments invested in renewing and developing infrastructure destroyed during the war. The post-war baby boom was also a significant stimulus for development in terms of the housing and services required by an expanding population. The housing boom in particular was dramatic in the post-war UK as demobilised servicemen needed to be accommodated along with their expanding families. Figure 5.1 demonstrates the dramatic expansion in all forms of housing throughout the immediate post-war years, reaching a peak in 1966–67. The feeling of optimism in the workforce at large was reinforced over successive years by consistently low inflation figures, particularly in the latter half of the 1950s. From 1957 to 1960 average inflation varied between a minimum of 0.6% up to a maximum of 3.7[1]. Low inflation allied to economic expansion and steady growth in wages saw a substantial rise in the real value of earnings throughout the 1950s.

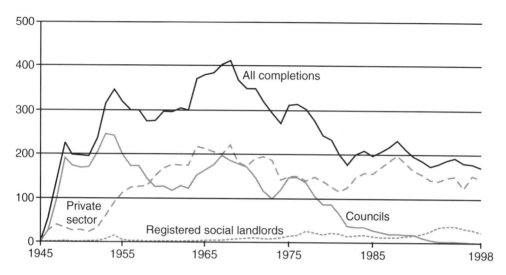

Fig. 5.1 Housing completions in post-war UK (adapted from Department of the Environment, Transport and the Regions; National Assembly for Wales; Scottish Executive *Social Trends Pocketbook*, 2000).

Job stability was also a major factor at the time, in that nationalised industries were the norm rather than the exception that they are today. In a very real sense workers expected a job for life rather than the significant turnover of workers seen today.

Socially, the 1950s were typified by a high degree of stability in households compared with the norm today. What we today refer to as 'traditional' values of the time are exemplified by the statistics associated with marriage; e.g. the total number of marriages recorded in 1960 amounted to some 400 000 while divorces numbered a mere 30 000[1]. This ratio of approximately 13:1 is not in itself that dramatic until it is contrasted with the current figure of some 3:1. Similarly, the vast majority of children born in the UK at the time were to married parents – single-parent families amounted to a mere 3% of all households in 1961.

In addition to the strongly delineated ideas of the 'norm' for families, there was also a strongly defined social structure at work. Trade unionism was extremely strong, with a significant number of closed shops, particularly in manufacturing and heavy industries. Trade unionism was also increasing in sectors not traditionally associated with such activity, particularly white-collar workers and healthcare, for example. Construction was similarly affected by the ability of unions to force stoppages, famously during the 1960s at the Barbican, for example. The average year at the time might see 250–300 stoppages per annum, involving up to a total of 50 000 men, losing literally hundreds of thousands of working days[2].

The zenith of this post-war expansion and social stability came in 1959–60. The 'never had it so good' phenomenon disappeared when, in 1961, balance of payments problems forced the Government to introduce an austerity programme. The resulting national reassessment of efficiency, value for money and cost effectiveness precipitated by the austerity programme focused attention on critical economic cost centres. Since the development of infrastructure is a significant expenditure for any Government, naturally enough construction became a focus for improving efficiency. The Emmerson report[3], which examined in great detail the problems of the construction industry, ultimately paved the way for further work on establishing methods and approaches to rectify the very obvious problems in the industry at that time. These were the prevailing conditions that formed the backdrop to the establishment of the Banwell Committee in the early 1960s.

5.2   Establishment of the Banwell Committee

In October 1962, and largely in response to the findings of the Emmerson report[3], Sir Harold Banwell was appointed to chair a committee investigating the state of the construction industry in the UK. The Banwell Committee was established by the Rt Hon. Geoffrey Rippon, then Minister for Public Works, who gave it a particularly wide-ranging brief. The Committee was intended to address two main objectives:

- to review current practice with regard to the placing and management of contracts for building and civil engineering work;
- to improve the overall performance of the industry by '[making] recommendations with a view to promoting efficiency and economy'[4].

The Committee assembled under Banwell was deliberately selected to be as broad based as possible in order to represent the diversity of professions in the industry. Indeed, the composition of the Committee included all the main professions in construction (surveyors, architects, engineers, etc.), construction industry suppliers, lawyers and clients. The main research tool used in order to garner opinion was that of an open-ended questionnaire distributed among a wide range of public bodies, professional institutions, trade associations and individual firms[4]. The questions posed allowed an extremely free range of responses. A typical phrasing of a question might be 'What are your experiences of the benefits and/or difficulties from the adoption of firm price contracts let by competitive tender?'. Given the difficulty of analysing responses to this type of question, the sum total of the input made by the Committee in its analysis should not be underestimated! Throughout the remainder of 1962 and

1963, the Committee met a total of 30 times, receiving written evidence from some 119 organisations and individuals. In addition to the written evidence, a further eight organisations and individuals were called to give oral evidence before the Committee.

5.3   Findings of the Banwell Committee

The findings of the Banwell report make fascinating reading, particularly in the light of the publication of the recent Latham[5] and Egan[6] reports into the state of the construction industry and how it should change in the future. Indeed, the most obvious points to note are that the main issues pertinent at the time have altered little. Furthermore, the recommendations made to solve the problems identified are, to a greater or lesser extent, quite similar to those propounded by the more recent commentators on the industry. The key observations and recommendations made by the Committee are broken down and briefly discussed, chapter by chapter, in the following sections.

5.3.1   *Chapter 1 – General observations*

One of the most interesting observations of the report is the allusion made to two prominent streams among members of the construction industry. On the positive side, Banwell noted the

> 'outstanding impression [of] an industry the progressive members of which are lively and full of new ideas, willing to experiment and not afraid to change their practices and procedures.'

However, on the negative side, the Committee also noted that this level of urgency, enthusiasm and purpose was not shared by the remainder of the industry. Indeed, the majority of the industry was governed by a set of inflexible contractual and professional conventions, to quote Banwell, 'designed in and for other days'.

The Committee goes on to recommend that trial and error should take place in order to establish the 'most suitable methods [to] secure efficiency and economy'. Progressive solutions to the problems of construction, what we today would refer to as best practice, should be coordinated and reviewed by a permanent secretariat within the Ministry of Public Building and Works (MPBW). This could be thought of as a very close analogy, with a similar mandate, to the Movement for Innovation (M4I) infrastructure set up as a result of the Egan report[6], and indeed part funded by the Government.

5.3.2   *Chapter 2 – The team in design and construction*

The Banwell Committee made some very pertinent observations with regard to the development of the construction team. Time, or more specifically the effective use of time, emerges as a key theme of the section dedicated to the operation of the construction team. There appeared to be a widespread inability of the construction team to dedicate enough time fully to plan out the programme of events for the project. The subsequent changes required during construction as a result of incomplete planning were noted as leading to additional costs through variations. Claims could be 'substantially reduced', the report noted, if participants could be made aware of the 'penalties of indecision and the cost of changes once the final plans have been agreed'. In order to rectify this deficiency, the main recommendation made by the Committee was that 'specialist consultants [should] be brought in at the earliest stage as full members of the design team'.

A further refinement of this 'up front' effort was also recommended in that the main contractor should be brought in to the construction team prior to design completion. The Banwell Committee justified this recommendation by the observation that

'Many [contractors] have developed highly specialised techniques in design and construction which can [be used] by the designer in formulating his scheme.'

To expect a contractor to be able to build a scheme without having some idea of, or indeed input into, how that scheme was developed was considered to be 'unreasonable'. Many of the difficulties highlighted were noted as resulting from a lack of mutual understanding between the disciplines. This observation appears, from a retrospective point of view, to be a veiled criticism of some professional institutions having a restricted scope to their training. The solution highlighted was that of common education among the professions, an issue that still regularly surfaces in university built environment departments to this day. Overall, if the Banwell Committee were to redraft their report today using contemporary phraseology, then it is likely that they would emphasise the fact that 'buildability' is enhanced by an 'integrated team' consisting of 'multiskilled, multifunctional' professionals.

5.3.3   *Chapter 3 – Appointing the contractor*

The report highlights the inefficiencies inherent in the 'open competition' approach to tendering, particularly among local authority and Government contracts. Banwell admonishes local authorities on their unwillingness to try 'unorthodox methods', i.e. limiting competition to save time and tendering costs, or even using negotiated contracts. This, Banwell notes, is often the result of 'rigid adherence to outmoded procedures'. As far as Banwell was concerned, open competition frequently led to situations where contractors would win a contract when they did not have the skills or capacity to fulfil that contract. Ultimately this would lead to cost escalation and delay. The key aspect in the selection and appointment of the contractor was that an element of preselection should take place before tenders are invited. Indeed, the report goes on to suggest that 'there will be occasions when competition may appropriately be limited further or even eliminated altogether'.

The report proceeds to encourage 'serial tenders' as a means of ensuring continuity of employment for contractors while building a skill base of expertise on certain works:

'banding together of those who have suitable work in prospect is [also] to be encouraged, which will allow construction to benefit from industrialisation and standardisation.'

For an individual reviewing the Banwell report in the light of recent developments in the construction industry, this sounds all too familiar. Preselection of contractors banding together in regular groups suggests a post-Banwell 'future' approach to construction remarkably similar to suggestions of strategic partnering[5] and the prime contracting route[6, 7].

5.3.4 Chapter 4 – Notes on procedures

Procedures adopted by both Government and local authorities come in for significant criticism. Of prime concern is again a more effective use of time, but on behalf of the main clients for construction. The report notes that approved contractor lists were not being kept up to date. This meant that historically inadequate contractors were still tendering for new work, while untried but potentially more efficient contractors did not get the opportunity to tender. The problem was further exacerbated by the fact that insufficient time was being allocated for the precontract evaluation of tenders, a particularly problematic scenario given that inefficient monitoring of the approved list necessitated the review of tenders from contractors that should not even be allowed to submit a bid.

Given the chronic inefficiencies that the Banwell Committee highlighted with regard to pre-tender administration, the post-tender procedures were also found to be inadequate. This is demonstrated by the fact that the Committee had to recommend the seemingly obvious requirements that 'the results of competitions should be notified promptly' and that 'adequate time must be allowed between the appointment of the contractor and the commencement of work'. Overall benefits to the process could be achieved, as noted in comments regarding the design team and the appointment of the contractor sections, by the earliest possible commitment of assets to the project.

5.3.5 Chapter 5 – Conditions of contract

After a review of the evidence presented to the Committee, the report recommended that

'a common form of contract for all construction work, covering England, Scotland and Wales, is both desirable and practicable.'

This recommendation is lent both weight and credibility by the fact that the Committee incorporated several representatives of the legal profession in its make-up. The report recognised that certain specialised situations may arise, such as the legal technicalities presented by projects in Scotland or from those commissioned by public bodies. However, it was recommended that these special cases be catered for by 'agreed additions or alternatives rather than by separate sets of conditions'.

Fundamentally, having a single form of contract that can be used for all work has significant advantages for all concerned. The main benefit is that there is not a requirement to 'reinvent the wheel' and spend substantial amounts of time deciding which of the myriad different contracts will best serve the project's specific circumstances. This approach has been recommended subsequently by Latham[5] specifically advocating the adoption of the new engineering contract (NEC) form for all construction works. One perceived weakness of the NEC itself was actually the name, which put some potential users off because they thought it to be substantially a civil engineering contract. Subsequently, the name and terms were slightly amended to engineering construction contract (ECC) so that 'construction' had a greater emphasis. The uptake of Banwell's advice, much as with the post-Latham construction industry, can most charitably be described as sporadic. Since the identical recommendation was given some three decades after the publication of the Banwell report, the recommendation seems to have been largely ignored by construction professionals.

5.3.6 Chapter 6 – Bill of quantities

The Committee recognised the overwhelming importance of an accurate and comprehensive bill of quantities in order to remove variations when contractors commence construction. The Committee recommended that improvements to the basic bill of quantities could be made by removing certain small labour-only items, for example, replacing them with a form of provisional sum. Further improvements in accuracy and the removal of variability can be made through the provision of improved supporting information for tenderers. This has been reflected by subsequent versions of the standard methods of measurement (SMM) for building and civil engineering works.

5.3.7   *Chapter 7 – Subcontractors*

The general consensus arising from the opinions presented to the Banwell Committee appear much the same as for the appointment of main contractors. Early involvement is highlighted as the key to success in general, and a high level of integration is required to get the most from subcontractors. The main contractor is perceived as needing to be responsible for the appointment of most subcontractors. Given that the main contractor should take responsibility for these subsidiary appointments, Banwell recommends that contractors should 'apply to the selection of subcontractors the same standards of fairness which they expect when they themselves are chosen'. Similarly, subcontractors 'should not be required to start work without a reasonable period of preparation'. These principles, essentially outlining how contractors can act responsibly and fairly when dealing with their subcontractors, are strongly reminiscent of the principles of partnering outlined by Latham[5] and reinforced, again, by Egan[6]. Fairness in dealing with subcontractors was not really an outcome that can ultimately be attributable to Banwell, since Latham was much more successful in this regard. Latham's recommendations[5] led to the Housing Grants Construction and Regeneration Act 1996, the significant contribution of which was to redress the balance between the dominance of the main contractors over subcontractors in the latter half of the twentieth century.

5.3.8   *Chapter 8 – Firm price contracts*

This chapter of the report examines extensively the need for clients and contractors to seek and establish firm price contracts as a mechanism to limit variations and risk. In essence this recommendation reflects the period of stability that the economy was enjoying at the time of the publication of the report. Inflation was low and therefore could be more easily factored into a contract. Later in the 1960s and running into the 1970s, inflation was markedly more volatile, and therefore firm price contracts went out of fashion quite rapidly in the industry, particularly for contractors.

In order to achieve a firm price contract, Banwell notes that, for any scheme, all the critical details need to be worked out, thereby leaving as little as possible to chance. Furthermore, when seeking firm price contracts, 'two years is not an excessive period for [their] duration'. To put this thinking into context, the Banwell Committee was deliberating and writing at a time when nationalised industries were still commonplace and therefore contracts and projects could be reasonably planned for some significant time ahead.

5.3.9   Chapter 9 – Payments, retentions and incentives

The Banwell Committee, in addition to the many other facets of poor management practices seen in the construction industry, also chose to comment upon the darker side of commercial management. Having first recognised that the payment of incentives is generally a good thing, noting that 'the loss of a bonus is a more effective penalty than a liquidated damages clause', the report goes on to examine the problem of payment and retention. Payments and retentions have long been recognised as being the source of conflict in the construction industry. Indeed, Chapter 9 of the report is actually the longest in the whole document, which in itself says much for the importance of the issue among those submitting evidence to the Committee. The general problem as seen by Banwell was that cash flow tended to be haphazard and often delayed, starting at the client end. The practice of building a fund of retained payments as contingency against failure of the contractor to deliver was particularly remarked upon.

In the polite phrasing of the time, Banwell explains:

> 'The operation of this system is not always smooth. Payments to the main contractor by the client are often slow and uneven, with consequential delays in payments to suppliers and subcontractors. This has an adverse effect on the efficiency and stability of the whole industry.'

Although the report goes on to explore in great detail the minutiae of how to improve the financial stability of the construction industry in general, one evident overarching theme was expressed succinctly by Banwell thus: 'What is needed [to improve cashflow] is an agreed procedure to ensure that payments are made regularly and promptly.'

This assertion is carried on through Banwell's recommendations. Firstly, public authorities in particular should not allow contractually due payments to be delayed by administrative procedures. Procedures should be arranged so as to permit financial obligations to be, as the report puts it, 'scrupulously honoured'. Where possible, additional provisions should be made for adequate payment for manufactured components in the process of development or production that are not ordinarily catered for in the standard valuation techniques in use at the time. If this is not done then the cost of manufacture of components will be borne by the manufacturer concerned for an indefinite period, which in turn puts unnecessary strain on the manufacturer and therefore the supply chain as a whole.

One suggested solution to the problem was noted by the Committee and included in the report. Specifically, certain Government departments had made arrangements with manufacturers to make interim payments for materials and components not yet on site, subject to the provision of a certificate of indemnity by the manufacturer concerned. A further recommendation notes that the use of maintenance, performance and tender bonds should be discouraged. Interestingly, Banwell goes on to recommend experimental projects both with and without using retentions and bonds, and 'the results of such an experiment should be made public'.

The general feeling that comes from an analysis of this chapter is that payments and retentions needed to be sorted as a whole throughout the industry. The justification for this is that there is a mutual dependency through suppliers and consumers to create cashflow in the construction process. This sounds remarkably like an exhortation to start actively managing the supply chain for the benefit of all – another echo of Egan[6]. Indeed, the suggestion that is made to conduct experimental projects to establish the veracity of new approaches, what might be termed establishing best practice, is reminiscent of the building down barriers initiative under the M4I scheme[7].

5.3.10   Chapter 10 – Scotland

No fundamental differences were identified within the nature and methods of construction taking place in Scotland. Indeed, as previously noted with regard to the recommendation to use a standard form of contract, all recommendations can be applied equally to both Scotland and the remainder of the UK. The only major recommendation made by the Banwell report that is specific to Scotland is that, in any future discussions on industry change in general, Scottish organisations should be represented. This recommendation seems somewhat condescending in the light of a post-devolution Scotland. However, it does actually say a great deal about the London/England emphasis of the various professional institutions at that time. Indeed, the Banwell report text actually states that

'steps must be taken [to] eliminate divergences of practice [and] for this reason, we trust that Scottish interests will be represented in the consultations [taking] place [following] this report.'

The implication here is that ordinarily no such representation would be included in the remit of further discussion!

5.4   Contemporary view of the Banwell report

The general consensus from contemporary reviewers of the findings of the Banwell Committee is that they contained much the same material, covering the same ground and reaching very similar recommendations, as both the

Emmerson[3] and Simon[8] reports. However, this should not be broadly taken as a criticism of the report – far from it. Indeed, as *The Builder*[9] noted very positively in its editorial comment on the report, 'reiteration is essential to education and to progress'. Each of the fundamental issues raised by the Banwell Committee was seized upon to a greater or lesser extent by the various interest groups of the time.

The Royal Institute of British Architects (RIBA) was reported to support two critical contentions within the Banwell report. The first was the report's emphasis on complete definition 'up front' in the project in order to allow a full process of planning. The second was that there should be increased flexibility in contractual procedures, a positive benefit of which was cited as being that 'the contractor could be appointed early in the construction process'[10]. Interestingly, in this cited article in *The Builder*[10], the RIBA makes some interesting assertions regarding the pre-eminence of the architectural profession in the construction industry. In the article, the RIBA observed that

> 'Safeguards would have to ensure that architects who work for contracting companies do not purport to offer a professional service to their customers'

which is something of a harbinger of the lack of regard sometimes shown by some architects for their contemporaries who work for a contractor!

As with the more recent implementation issues surrounding the Latham[5] and Egan[6] reports, one overarching theme of contemporary comments on the Banwell report is that many of the recommendations are 'already being dealt with'. However, particular support was given to the recommendation to adopt a strategy of selective tendering[9], with an initial report that Model Standing Orders be investigated by the Ministry of Housing and Local Government[11].

The Minister for Public Building and Works, Geoffrey Rippon, who originally commissioned the Banwell Committee, held a press conference to announce the findings[12]. During this press conference, the Minister discussed and broadly accepted all the recommendations made by the Committee with only some very minor reservations. Mr Rippon announced that the Government would 'take the lead' on the implementation through the provision of all secretarial and financial assistance required to implement the recommendations. At the time the recommendation to set up a new secretariat was put on hold in favour of working through the normal ministerial channels, allied to the National Joint Consultative Committee (NJCC).

For some commentators on the effect of the Banwell report, there was not sufficient depth of consultation. One plaintive reader of *The Builder*[13] noted with alarm that the Working Party responsible for drawing up 'standard forms of tender and acceptance' did not actually include any representatives of subcontractors. Indeed, although the final forms created by the working group were 'imminent', the Federation of Associations of Specialists and

Subcontractors, i.e. those most likely to be affected by their implementation, had not been asked to agree their final form. The nature and tone of what was written at the time prompted the Government of the day to attempt to broaden the impact of the Banwell recommendations for the benefit of the industry and the country in general. This resulted in the establishment of further working groups to gauge the 'progress and adequacy' of the measures adopted or contemplated in the light of the Banwell report recommendations.

5.5   Implementation of the Banwell recommendations

The degree to which, along with the methods by which, the recommendations of the Banwell Committee could be implemented by the construction industry remained something of a moot point for some time. Initial change came about relatively quickly among certain disciplines. For example, following the Banwell report, 1966 saw the implementation across England, Scotland and Wales of the standard method of measurement version 5[14] for quantity surveying. However, little evidence can be found to support the contention that wholesale changes were occurring.

Consequently, in July 1965 the Economic Development Committee for Building (EDCB) appointed a subsidiary working party on the implementation of the Banwell report findings specifically within the building sector. The Economic Development Committee for Civil Engineering (EDCCE) followed suit in October 1965 by commissioning a Banwell implementation working party in the civil engineering sector. Both working parties subsequently reported back and the findings were published[15, 16].

The two reports made various pertinent observations on the uptake of the Banwell report, but do not diverge substantially from the thrust of the original Committee findings. What both reports did do was to suggest procedural refinements in order to expedite the recommendations of the Banwell Committee. For example, the EDCB report[15] strongly supported an enhanced role for both the NJCC as well as an expanded and enhanced membership of the Joint Contracts Tribunal (JCT). These modifications were anticipated more readily to create a single 'best way' of structuring contracts that was acceptable to all levels within the industry. Similarly, all the professional institutions [RIBA, the Royal Institution of Chartered Surveyors (RICS), etc.] and the Government's MPBW were recommended to become proactive, issuing either professional rules or codes of practice to reflect the changes recommended by Banwell.

Overall, the EDCB[15] and the EDCCE[16] reports noted that Banwell-inspired change was underway but that it would take some time to permeate the industry fully. Similarly, the amendments and refinements advocated by both these subsidiary reports needed to be followed through before a full and complete review of the industry was undertaken. This effectively laid the field open for further refinement and improvement in the coming years.

What is particularly interesting is that, although changes were recommended by Banwell[4] and refinements were advocated[15, 16], none of these reports actually sets out targets for when all the changes were to be completed. Similarly, no key criteria were put in place to measure any changes that did occur. In essence, therefore, if change were effected it would be extremely difficult to define whether or not the systemic change had actually created positive improvement. This deficit was addressed in the recent Egan report[6] which advocated the use of key performance indicators for all processes in order to measure change.

5.6 Banwell: catalyst for change or overtaken by events?

A review of the literature available from the period shows that the Banwell report was as significant in its time as the Latham[5] and Egan[6] reports were in the 1990s. A significant amount was written in contemporary journals and provoked lengthy debate, but its impact waned over time. Several factors contributed to this.

5.6.1 Activity levels in the industry

The 1960s demonstrated relatively even demand for construction of housing and infrastructure which allowed longer-term planning for companies in the industry. However, as early as 1965, Lewis[17] had noted that the prosperous years in the construction industry were likely to disappear before the end of the 1960s. Subsequently, Ball[18] reported that, from the mid-1960s to the 1980s, there was a systematic decline in real-terms output by the industry, associated with periods of 'general economic crisis'. There appears to have been a combined effect of reduced construction funding from central and local government, as well as the more generalised effects of the 30 year economic cycle in the UK moving towards recession. These macro trends seem likely to have shifted the emphasis for most companies away from trying to make long-term process improvements towards economic survival.

5.6.2 Change in funding for the industry

Following on, the 1960s and early 1970s saw a shift in the way in which construction companies were in themselves funded. Increasingly, the sector saw the emergence of public limited companies (PLCs) instead of family owned and run enterprises. As more private money moved into the construction arena, so pressure gradually increased for construction to demonstrate a return on investment (ROI). In order to achieve improved ROI, the industry substantially restructured itself, moving from employing large numbers of workers and skills 'in house' to an almost entirely outsourced/subcontracted organisational structure. Ball[16] reports that, during the 1970s, the ROI for the construction industry rose from 15 to 18% as a result of some of these strategies. Subsequently, the ROI for the industry has been in decline, while other industries have remained fairly constant. Arguably, this could imply that all the 'improvements' possible in ROI from outsourcing have now been accrued, while pressure on prices has increased. Either way, the substantial structural changes undertaken by the construction industry have overtaken many of the changes and improvements recommended by Banwell, particularly with regard to the terms and conditions of contract.

5.7 References

1   Office of National Statistics (2001) *Social Trends Pocketbook*. HMSO, London.
2   Finlayson, J. W. (1971) Manpower and motivation. In: *Construction Management in Principle and Practice* (ed. E. F. L. Brech), pp. 549–651. Longman.
3   Emmerson, Sir Harold (1962) *Survey of Problems Before the Construction Industries*. HMSO, London.
4   Banwell, Sir Harold (1964) *The Placing and Management of Contracts for Building and Civil Engineering Work*. HMSO, London
5   Latham, Sir Michael (1994) *Constructing the Team*. Report from the Joint Review of Procurement and Contractual Arrangements in the UK Construction Industry.
6   Egan, Sir John (1998) *Rethinking Construction*. Report of the Construction Task Force, HMSO, London.
7   Nicolini, D., Holti, R. & Smalley, M. (2001) Integrating project activities: the theory and practice of managing the supply chain through clusters. *Construction Management and Economics*, 19, 37–47.
8   Simon, Sir Ernest (1944) *The Placing and Management of Building Contracts*. HMSO, London.
9   *The Builder* (1964) The Banwell Committee reports. *The Builder*, CCVI(6311), 1 May, 901.
10   *The Builder* (1964) RIBA's comments on Banwell report. *The Builder*, CCVI(6311), 1 May, 989.
11   *The Builder* (1964) The Banwell report on contracts; NJCC's comments. *The Builder*, CCVI(6313), 29 May, 1127.
12   *The Builder* (1964) Mr Rippon accepts Committee's recommendations. *The Builder*, CCVI(6311), 1 May, 922.
13   *The Builder* (1964) NJCC on the Banwell report. *The Builder*, CCVI(6315), 12 June, 1135.
14   Francis, R. H. (1963) *Standard Method of Measurement*, 5th edn. Standard Joint Committee, Royal Institution of Chartered Surveyors and National Federation of Building Trades Employers.
15   EDCB (1967) *Action on the Banwell Report*. HMSO, London.
16   EDCCE (1967) *Contracting in Civil Engineering Since Banwell*. HMSO, London.
17   Lewis, J. P. (1965) *Building Cycles and Britain's Economic Growth*. Macmillan, London.
18   Ball, M. (1988) *Rebuilding Construction: Economic Change in the British Construction Industry*. Routledge, London.

Chapter 6

Tavistock Studies into the Building Industry: *Communications in the Building Industry* (1965) and *Interdependence and Uncertainty* (1996)

David Boyd & Alan Wild

6.1   Introduction

This chapter is a tale of two interim reports that came out of a unique research project; as the 1965 report cover stated:

> 'the study breaks new ground in two important aspects: it is the first example of an industry as a whole inviting research into its own operations; and it is the first application of the combined resources of operational research and the social sciences to a research project.'

Apart from this, the research investigated the detailed operations of the industry, unlike previous studies that saw only a macroeconomic analysis, and produced some challenging and original observations. The novelty of the approach, the complexity of the subject and the abstract language of the analysis may have led to the work having little direct effect, but it can be seen as the precursor to many later studies of the industry, and also some of their solutions are those currently being used to change the industry.

The Building Industry Communications Research Project (BICRP), as the project became known, was instigated by the National Joint Consultative Council of Architects, Quantity Surveyors and Builders (NJCC). It was not initially commissioned by the Government. The Tavistock Institute received the commission on 12 October 1962 for a pilot study for completion by December 1962. It was conducted by Gurth Higgin, a psychologist of the Tavistock Institute, and Neil Jessop, a mathematician and statistician of the Operational Research Society (ORS). Following the positive reception of the pilot report at a conference in March 1963, much negotiation took place to establish a 'collaborative research institution' in order to continue the research. Using funds drawn equally from the industry and Government, a 2 year project was started in January 1964. The research team, at the now fully constituted Institute of Operational Research at the Tavistock Institute, was led by Higgin and Jessop but was enhanced by John Luckman, John Stringer and Don Bryant, all on the operational research (OR) side. This team's work continued for over a year until funds became exhausted. The uncertain atmosphere in the industry, administrative difficulties in the project and the publication of the 1964 Woodbine Parish[1] report on a similar theme of building research and information services prevented more money being accessed and the work continuing. Also, the Banwell report[2] moved thinking away from understanding the industry back to Government agendas of comprehensive solutions with a focus on contracts.

In fact, the report *Communications in the Building Industry* was not fully published and distributed generally until 1965 owing to the conflicts within the steering committee and research team. The report *Interdependence and Uncertainty* was considered an interim document to record the work undertaken before the funds ran out and leave a bridge to further work. This latter report by Charles Crichton, an architect and journalist from outside the research team[3], demonstrates something of the conflict that the team and the steering committee had fallen into. Clearly, the two reports share a common historical and production context, but they both delivered unique, although connected, findings.

6.2   Britain in the early 1960s

The early 1960s saw a positive feeling in the country, with growing affluence and a belief in a better future with major social change[4]. This feeling disguised a poor economic position for the country both in relation to the rest of the world and in the fixed and social infrastructure. It is this paradox between belief and reality that governed actions and experiences through this period.

As victors in World War II, when, as a country, people had worked together through the fears and agonies of war, there was a belief in a rightful position in the world and that citizens deserved a reward from action. This caused Britain to maintain expensive activities overseas, including a nuclear capability, while increasing wages and changing social structures. In particular, home ownership doubled to nearly 40% from the early 1950s to the 1960s, and there was a major programme of slum clearance. The reality was that the war had effectively bankrupted Britain, with a major reduction in gold and currency reserves, while post-war reconstruction was based on loans from the USA and Canada on what was regarded by many as extremely unfavourable terms[5]. Britain's share in world trade declined as pre-war competitors recovered. Thus, in 1950 Britain had 25% of world trade, but by 1962 this had fallen to 15%, with the difference taken up by Germany and Japan[5]. The relatively positive climate of the 1950s started to go wrong with the first recession of 1961. There ensued for the next 10 years a number of balance of payment crises when the phrase 'stop–go' economy was coined. This required cuts in public expenditure and a squeeze on credit, both important aspects for construction. Much of the construction work of the post-war era was for the state. This had a heavy influence on the nature of work, the approach to this work and ultimately on the volatility of the quantity of work. Because of the backlog of work after World War II, construction output rose steadily after restrictions were lifted in 1954[6] so that by 1963 output was about £11 000 million (1975 prices), almost 50% more than in 1955. In order to meet the reconstruction demand at the right price, improving productivity became a major issue, and this concept fitted with an economist's view of the world. Productivity could be improved with better training and the use of a more skilled workforce. Finally, production technology could be improved, and much work was being done on standardisation, prefabrication and the use of industrialised building systems. However, it was the volatility of demand that made it very difficult to match supply and demand, so that organisations were forced to structure themselves and operate around extreme flexibility with less attention being paid to project efficiency[7].

The early 1960s were a time when a Conservative Government had been in power for 10 years and had experienced political scandals, failed foreign policies and failed technological developments. The defeat of the Conservative Government in 1964 by Labour under Harold Wilson was with a small majority. This made the new Government's job of balancing the conflicting issues between social, business and economic agendas very difficult, and, in particular, desperate efforts were made to avoid devaluation. The economy worsened, but the Government's attempt to gain a larger majority at the general election in 1966 was successful. The arrival of a Labour Government was viewed by the construction industry with great suspicion; issues such as nationalisation of the industry and even of land were being discussed. As important to the despair within the industry was the continuing stop–go nature of public projects. Attitudes to one nation working together which had survived since the war were breaking down.

Since World War II there had been a strong belief in technology both as an economic advancer and as a symbol of a positive future. Britain's world role caused it to develop many advanced technologies from aerospace to defence, although several were not entirely successful. Together, this technological and economic failure caused a review of education and of industry in the early 1960s. There was a strong belief that what the economy and industry needed was a more rational approach to thinking about the world. Some of the criticism of economic performance was levelled at management which did not attract enough high-calibre people to industry who had sufficient experience of the shop floor. Thus, the context of the Tavistock studies was an egalitarian outlook that crossed previously quite rigid class and professional boundaries, a strong belief in technology for new products, new processes and new organisations and, finally, a rational planning and decision-making process as a modern model of action.

6.3   Building Industry Communications Research Project

Officially, the BICRP emerged from a decision of the NJCC, but the Emmerson report[8] had clearly provoked reaction from the industry and provided an emphasis for the direction of the study towards communications. The commissioning of the Tavistock Institute to undertake the research was partly because of its neutrality, its experience in industrial research and its ability to offer scientific solutions from operational research. The Tavistock Institute for Human Relations is a unique institution both in its perspective and in its constitution. It developed out of the Tavistock Clinic set up in the 1920s, which gave it a psychological perspective on the world. By the early 1960s the Tavistock had a track record of engaging with a number of complex industries, notably coal mining, engineering, farming and the chemical industry. Its association with the ORS presented an attractive package of soft and hard skills that promised effective solutions. From World War II, OR offered rational and numerical modelling techniques to enable the rapid and effective improvement of production processes.

The project was given credibility by a group of six establishment trustees including Sir William Holford, Sir Hugh Beaver, Sir Harold Emmerson, Sir Harry Pilkington, Mr D.E. Woodbine Parish and Mr Cyril Sweett. The fortuitous links between Emmerson, Beaver and the Tavistock must be acknowledged, and also how these implied that this innovatory and risky project was in safe establishment hands. Sir Hugh Beaver, who was Chairman of the Tavistock and Chairman of the new National Economic Development Council (NEDC), was also Emmerson's predecessor as Chief Secretary at the Ministry of Public Buildings and Works. As well as this, a steering committee of 20 prominent people from the industry assisted with the study under the chairmanship of Lord James of Rusholme. This steering committee, in the egalitarian view of the time, contained representatives from the professions, Government research, main contractors, specialist subcontractors and a trade unionist.

6.4   Executive summary of *Communications in the Building Industry*

The aim of this pilot study was to define the scope and cost of a more major project. In the research proposal to the NJCC of August 1962, the approach (and analytical outcome) was clearly declared: that the Tavistock/ORS would consider the *building process* based on the client's building needs, determining the *resource controller's* role in communications (pp. 88–90). The result was a restatement of many ideas already held by the industry but cohesively based around the Tavistock's preconceived framework for understanding organisations. Being a pilot study and also being designed to stimulate discussion, there was little attempt in the report at ensuring a complete logical exposition. The contents of the report are not introduced and so there is little explanation of its overall structure, which includes five sections and two appendices. The main section, which is Section 2, consists of 20 pages of detailed general process mapping of operations, where the authors try to identify communication needs and problems at different stages in a project. From this the researchers determined that operations depend on the roles and relationships established in projects (this is analysed in Section 3) and that improvements could be made through the use of some more formalised OR techniques (this is discussed in Section 4). The final section forms a set of five questions for further study and lays out recommendations and requirements for this study. The first appendix simply records the NJCC requirements for a research study. The second appendix details a piece of unfinished research based on a questionnaire on how different members of the building team rate the social status and the importance of the contribution of other members of the team. The results of this are presented, but the authors refuse to make any comment or deductions from this.

6.4.1 Construction process mapping

Communications are divided into eight phases:

- Phase 0 – client deciding to build;
- Phase 1 – client consulting the building team sponsor;
- Phase 2 – sponsor investigating and preparing the brief;
- Phase 3 – preparing and gaining the client's acceptance for sketch plans;
- Phase 4 – preparing contract documents, obtaining final approvals;
- Phase 5 – preparing and agreeing the contract, setting up a construction team;
- Phase 6 – construction to completion;
- Phase 7 – handing over and settling the final account.

Two themes appear in all phases: firstly there is considerable variety and confusion in the arrangements, and secondly there is a distinction between formal and informal communications. This is the first time that the informal aspects of construction were acknowledged. The first four phases are all concerned with the client and the relationship between the client and the industry. The authors draw a distinction between the level of sophistication of the client according to understanding and experience of the building process, which affects the way they make approaches and decisions. It is this that determines the long-term pattern of communications and therefore the progress and the success of the project. The client selects the 'sponsor' who is the team member first approached to discuss building needs; this may be the architect, quantity surveyor or builder, depending on the client's viewpoint. From these informal yet strategic decisions, communications proliferate quickly, but the sponsor has a particular power in the ensuing process. Naive clients and corporate public bodies have difficulty in defining and communicating their needs, and the sponsor often feels frustration from an inability to help. However, sponsors work from within their professional roles which limits their appreciation, constrains their solutions and polarises criteria of success to their special concerns. This all works against the client's needs and causes conflicts within the building team. The authors suggest that social science tools for exploring needs and communicating them within groups may assist here.

All the above are strategic processes that set the structural and relationship framework of the project. Once the client's acceptance of the sketch plans is required, the project has moved to a tactical stage with an enlarged team composed of numerous specialists. The authors present one of the most significant observations of the report that

'the problem arises of the need for *continuous intercommunications* between interdependent activities independently carried out.'

This induces a shifting ground for decision-making, such that

'by the time the consultant has arrived at his answer and communicated back, the situation described in the original communications may no longer be valid. Thus his reply is no longer relevant.'

Time and cost constraints make errors as a result of this more likely, and tendering procedures induce opportunistic behaviour and problems later in the project. The authors see that formal communications are 'not used as fully as they might be', but, anyway, they believe that the available techniques do not 'ease the problem of constant intercommunications between concurrently developing design activities'. They call for research into design decision analysis that could clarify, simplify and make more relevant these processes to achieve the necessary optimisation from the 'balance of suboptimal solutions'.

The authors see the preparation of contract documents as the transition from the professional to the commercial arena, reflecting the split between design and construction. This is the second major observation of the report where a highly abstract interpretation is provided. They are drawn to the role of the quantity surveyor (QS) and the way that the bill of quantities (BoQ) should 'allow the passage of the project from the one arena to the other'. Having set up this important role for the BoQ, they dash it by revealing that the BoQ can never fully describe the building structure that will finally emerge. They conclude that

> 'the bill of quantity is a hypothetical construct and not necessarily a fully accurate description of reality, despite its detail.'

They feel that understanding this could lead to 'less tedious, more economical and equally effective techniques ...for achieving the same ends', that is estimating could link to operations.

As the project moves to the construction phase, the authors see this as mainly a routine process with some traditional rituals in letting, receiving and deciding contracts that settles relative positions of financial interest and responsibility. The contractor takes on the role of the sponsor of the construction team, seeking to optimise the mix of resources. However, during construction, the problems of roles, inconsistencies in documentation and variation orders come to the fore. If a subcontractor has been involved in design then he has an alignment with the architect outside his contractual relationship with the main contractor. This causes role and managerial conflict for all parties. Inconsistencies between drawings, bills and specifications may emerge, and this can allow parties to take opportunities for claims for extra money; thus, there is a vested interest in faulty communications. The same thing can apply to variation orders, where communications will almost invariably lead to difficulties for somebody. The type of contract is important to the level of self-interest resulting from faulty communications, such that the authors believe that there is some advantage in package deals.

As a project is completed and final accounts settled, the authors see that much of the acrimonious argument is due to the accumulation of the results of inadequate communications throughout the whole building process. The two overall conclusions from the process mapping were firstly that 'the main factor lying behind communications difficulties is the nature of the relationships between the communicators', and secondly the lack of valid information 'about just what job any communication is supposed to do'. The next two sections, one on roles and the other on operational research, attempt to address these issues.

6.4.2   Roles and relationships

The authors, having identified that there was considerable variety and confusion in roles, set out to analyse how this affected communications. They presented a social history of the industry and showed how roles had developed from the wider social and economic pressures. Thus, the needs for construction had been driven by external events, for example the rapid rebuilding after the Fire of London in the seventeenth century, the demands of the industrial revolution and the Napoleonic wars in the eighteenth and nineteenth centuries; technical developments, population growth and urbanisation in the nineteenth century and, finally, the war reconstruction in the twentieth century. In response to this, construction had first moved from a guild structure to a trade and entrepreneurial contracting system mirroring commercial developments in the rest of society. At the same time, the role of the architect had emerged, with design separated out under the aristocratic patronage of London clients in the 1780s, followed by the role of the quantity surveyor to account for costs within a more fragmented delivery system. The role of the subcontractor had emerged as developments in technology required more specialisation, with ensuing problems of control and accountability indicating a need for explicit management.

Throughout these 300 years of development, the industry has been periodically accused of failing to meet client needs, most commonly concerning cost and time. In response, starting from the Napoleonic wars, the Government has felt it needed to intervene both for national economic efficiency and also as a major client. This involved the creation of special legislation to prescribe processes and responsibilities and the setting of roles and relationships within a contractual framework. This gave roles a new meaning and status within the law, society and the industry, which introduced a new set of problems around disputes and litigation.

The authors returned to the Tavistock's psychological perspective on organisations in order to present another original observation. They saw roles as confused and unstable, varying significantly across different projects, which induced 'a general anxiety among all concerned'. Thus:

> 'In this situation it is not surprising that relations between the various parties should often be strained and that tensions should exist .. As a result there is an understandable defensiveness on the part of everybody, particularly when entering a new relationship. In the absence of generally agreed rules for the relationship game, every man wants to ensure he is not a losing party. Natural developments of this are, on the one hand, the offensive/defensive stance that corporate bodies representing the different roles tend to take up with each other, and on the other, a formal amiability that denies the underlying tensions' (p. 52).

This allowed them to conclude that

> 'Any lack of cohesion and coordination is less the result of ill-will or malignancy on the part of any group or groups, but more the result of forces beyond the control of any individual or group and which are affecting all' (p. 53).

and to suggest that solutions are not simple but require a full understanding by everyone of the situations. It is from this that it would be possible to agree 'redefinitions of responsibilities and the stabilising of relationships which would allow for greater joint effectiveness'. The authors suggested that OR could assist this and use the next section to explore possibilities.

6.4.3 Operational research

The authors defined OR as a scientific approach to complex problems in the management of large systems of men, machines, materials and money. Its aim was not merely to predict the optimum behaviour but to forecast, by mathematical techniques, the behaviour of systems conceived in entirely different ways. They outlined the critical path method to assist the unfamiliar audience of the time and attempted to demonstrate how it could be used for coordination. However, in a very perceptive way they understood the limitations of this as being the complication of the construction controller's resource allocation problem by competitive demands and also the effects of accidental factors, e.g. weather, and variations in the time taken to complete the same job. Their OR solutions to this were queuing theory and simulation modelling although they did not explore these further. They saw the need for fundamental research into OR methods

> 'to quantify roles in terms of their relative effects on the interdependent operations and the inevitable relationships between roles this gives rise to.'

6.4.4 Recommendations

In conclusion, the authors identified five problems that needed to be addressed:

- How can the industry help prospective clients understand what can be done for them?
- How can the industry help clients to have their needs met during the building process?
- How is the design team built up and how does it communicate to create a total design?
- What minimum information is required in contracts so that those concerned know what is expected of them?
- What is the nature of communications in the construction team to ensure efficient construction control?

They recommended that, through a study of 'the chain of interdependent activities', the roles and 'distinctive competences' of the resource controllers be clarified. Through this understanding, a reordering of the divisions of responsibility might be possible that could reconcile more effectively the technical and organisational interdependences, and this should be studied in a real-life laboratory where a series of projects could be undertaken under conditions of experimental protection. They saw this as requiring the cooperation and sponsorship of the industry as a whole, with Government support.

6.5   Executive summary of *Interdependence and Uncertainty*

The report *Interdependence and Uncertainty* was considered an interim document to record the work undertaken and leave a bridge to further work. Again being an incomplete piece of work, like the pilot study, the report lacked a finished cohesion. There were three parts plus appendices. The first part concentrated on the OR issues, demonstrating the dominance of this thinking. The second part extended the sociological analysis introduced in the pilot study by explicitly using a socio-technical systems approach. The final part presented a short discussion on possible improvements and future research The appendices gave some extracts from the case studies in order to demonstrate the roots and effects of uncertainties within the construction process.

6.5.1   *Operational research*

The notion of *interdependence* had run through the researchers thinking right from the beginning. This was still paramount in their analysis and the central focus of the OR solutions. However, the notion of *uncertainty* was new. The essence of this was addressed in the pilot study but was expressed there as variety and confusion in roles, the inadequacy of documentation and the disjointed progress of projects. The authors noted a number of sources of uncertainty from outside the project, such as demands on resources from other projects, and from within the project, such as the changing participants and relationships as the project develops. These uncertainties induce anxiety in participants which prevents them from acting in the project's, and their own, best interest. The author saw ways of lessening uncertainty through less adversarial procurement methods which would enable earlier involvement and from proper and consistent use of OR tools.

The OR team devised AIDA (Analysis of Interconnected Decision Areas) as a method of finding technical solutions in these environments. The team established that the construction process was using a sequential decision-making approach to what was a highly interdependent situation. They felt that, if the form of interrelatedness of all decisions were described at the beginning of the decision-making process, then 'consistent decisions could be made more swiftly and meaningfully'. AIDA attempts to do this.

AIDA breaks down a technical problem into 'decision areas', for example the type of roof. Having identified all the decision areas, it is possible to identify the interdependent relationships between them and to show them diagrammatically on a 'strategy graph'. This representation allows the set of compatible options to be identified (in fact, AIDA identifies incompatible options), and this is displayed in an 'option graph'. All possible solutions can then be costed in order to identify a best solution. The report revealed that the project team had spent much time in doing this for a house and even then indicated that larger projects would require a computer.

6.5.2 *Sociological perspective*

The sociological chapter developed the problem of the client introduced in the pilot study and then engaged in a socio-technical systems analysis of the competitive tendering process. The socio-technical analysis modelled the construction process as five closely linked subsystems: operations, resource controllers, formal controls (directive function), informal controls (adaptive controls) and social and personal relationships. The client system is not included as it has interests quite separate from the project's and industry's concerns. This was a profound analysis that brought together much of the original thinking, although it was extremely abstract and difficult to apply to practice.

The operations subsystem considers the purely technical requirement of the process, i.e. the bringing together of materials and information to produce a working building. This exhibits interdependence which makes the phasing of technical decisions crucial. The specialisation of function adds to the problem and induces particular requirements for control. The other subsystems all involve human activities. The first of these, the resource controllers subsystem, which includes functional roles such as that of architects, controls the way in which people, materials and information are brought into the system of operations. These roles have developed from needs in much earlier times and have an inertia which means that the roles are entrenched and do not meet the needs of the system of operations. 'Fortunately, people can and do cheat in carrying out formal roles, so collaboration of a more practical kind is possible …informally'.

The formal controls subsystem is set around the resource controllers. It is this subsystem that is described, written down and taught as the way construction occurs. This subsystem is characterised by independence of function and by 'sequential finality' which assumes that lines can be drawn marking the beginning and completion of any activity. Such fragmented organisational form and procedure are not suited to the interdependent and uncertain situations found in building. Thus, both the arrangement of resource controllers and the formal controls require a system of informal controls in order to make projects work.

The largest section of the socio-technical analysis is the discussion of the informal controls subsystem. This is subtitled 'adaptive function', indicating that it is how the overall system responds to circumstances. The author reveals that much of this behaviour is quite conscious but hidden and not acknowledged. However

> 'the informal procedures seem to produce more realistic phasing of decisions, more continuous application of control functions and more realistic flexibility in face of the inevitable uncertainties that all must accept.'

The principal reason that the formal system does not work is its assumption that information, whether brief, drawings or bills of quantity can be complete before construction. It is the lack of complete knowledge by all parties that causes decisions to be modified as events unfold. In particular, 'the builder's price and client's acceptance of a competitive tender must always be an act of faith'. Various informal insurances are put in place by parties to protect themselves from this, such as builders overpricing certain items and QSs hiding contingencies in items within the bill. Many of these actions are undertaken collusively in acceptance of the unreality of the formal system. It is this need for the informal system and the fact that it is hidden and not open to discussion that prevent scientific management techniques from being effective. Thus, the informal system has the disadvantage that there are very short forward planning cycles of days which induces a climate of endemic crisis. People either become skilled in surviving such a climate or leave the industry.

The final subsystem, that of social and personal relationships, is where the stresses of the failed formal system and the hidden collusive informal system are worked out. Projects commence with faith in an advantageous outcome, but, as more problems occur, the informal system cannot cope. At this point, individuals are forced to protect themselves from fault or try to place fault on others which creates acrimonious relationships without solving any problem. This hiding of problems and protectionism is exacerbated by competitive tendering, both of the main contractors and subcontractors, which encourages lying to gain a job. As people meet

> 'all are imprisoned by their collusive acceptance of unreal, independent accountability for parts of an interdependent responsibility.'

The reasons for these situations are commonly seen as 'incompetence, laziness or financial greed of others' rather than as a failure of the system as the author believes is the case. This apprehension makes the situation worse.

6.5.3   *Suggestions for future research*

The final part considers possible improvements, acknowledging that the criterion for these cannot be easily established as members of the project team hold diverse values. The author believes that the main reason for inefficiency is systemic uncertainty, creating an environment of conflict. He states that natural evolution would be the proper vehicle for change but that this is slow and makeshift. A more directed effort could come from the understanding delivered by future research. This, he believes, should look at radical changes to the overall organisation of projects and the industry so that a much wider coordination of control is achieved. This could be analysed by such techniques as AIDA. However, the starting point must be the customers both in the sense of understanding their particular system and of delivering the measure against which successful change can be assessed. In conclusion, he sees the 'collaborative leadership for change' displayed by the BICRP as being a model for future action, but this requires a new focus.

6.6   Impact of the reports

The BICRP did not have a direct influence on the industry, but it did create a new set of theories not just about the industry but also about complex, interdependent and uncertain situations in general. These more general theories have survived.

The research, by shifting the focus on construction away from contracts to uncertainty and interdependence and to the analysis of the roles of resource controllers, had for the first time developed a real appreciation of the meaning of fragmentation in an industry. This was published later by Stringer[9], a research team member, under the concept of temporary multiorganisation, for the more general context of intercorporate relationships. The many decision-makers from different organisations, with their own affiliations, goals and values, influence resource allocation within shared projects and induce further uncertainties. In order to manage and influence such situations, values need to be clarified and mapped across the political and technical agendas, with the result that decision-making, problem-solving and planning need to become decentralised and collaborative activities.

Taking up a leftover problem from the BICRP, Bryant *et al.*[10] started working on a deeper understanding of client systems. This was later developed under a Science and Engineering Research Council (SERC) grant by Cherns & Bryant[11]. Clients' systems exhibit conflicts internally between different interest groups because of competition for scarce investment resources which forces people to misconstruct risks and costs. Power accrues to the 'winners' who become hostages to fortune in relation to the success of the project. The hostages include personal reputations and professional and functional interests that focus differentially on the limits of time, cost, quality etc. Therefore, the client is not unitary and prior events within the client system affect the current conduct of the project.

A wider analysis of uncertainty and a method for handling it in groups with conflict was brought together by Friend & Hickling[12] and became known as *strategic choice*. This work started with Neil Jessop in the late 1960s, who looked at strategic planning in the city of Coventry and identified three different uncertainties in situations as uncertainties about guiding values, uncertainties about the environment and uncertainties about choices. Each of these required different ways of working: clearer policies, deeper investigation or broader perspectives respectively. However, this analytical perspective was balanced by a highly participative and collaborative learning approach involving problem shaping, solution designing, solution comparing and solution choosing. It is the introduction of this balanced approach to problems involving both rational analysis and collaborative decision-making that must be regarded as a major success of the BICRP.

6.7   Discussion and critique

We believe that the BICRP was an exceptional research project for three reasons. It was unique in being commissioned by a cross-industry group as a cooperative enterprise rather than a Government report on the industry. It was unusual for that time in attempting a process analysis of project operations that bridged social and technical perspectives. Finally, it produced some insightful but controversial theories about the industry that acknowledged informal communications and revealed that the variety and confusion inherent in the systemic problems of the industry induced anxiety in participants, thus making it difficult to deal with value conflict. The larger team compared with that of the pilot study allowed a more sustained investigation into the industry and enabled the research into uncertainty and interdependence to clarify issues and arguments. The agenda became a search for rational analyses and tools for dealing with interdependent and uncertain situations. The outputs such as AIDA and the socio-technical analyses did not impress the industry because they were too abstract and basically unusable. *Interdependence and Uncertainty* was very academic and the longer-term results of this research are more evident in other fields. The team fragmented and worked on individual aspects of their perceptions but in more general circumstances.

6.7.1   *An industry report*

That it was not a Government report is significant because it gave the authors the ability to expose power interests and to challenge a wider perspective. It is this neutrality that was expected from the Tavistock, but the implications of this may not have been considered – it threatened power interests. Government reports are by their nature political, that is, they are a presentation for or against a line of thinking within a power context inside and outside the industry. Most often people in power are canvassed for their views, the assumption being that solutions are out there among these experienced people and it is possible just to combine a number of these perspectives in a rational way to meet the objective. Government reports deny both self-interest from powerful stakeholders and the existence of legitimate conflict. What was challenged in the report was this wider model/construct of the world and how experience both was generated by this view and generates this view.

6.7.2   *The nature of the industry: systemic conflict*

The BICRP was probably the first piece of research done on the structure and operations of the industry. The Tavistock had been developing an understanding of the operation of a number of complex industries in order to help them develop. They were part of the systems movement[13] and had made particular contributions in 'open systems analysis' where the interpenetration of external and internal problems was important and in particular explored wider social systems than that of the single organisation on a much longer time horizon. Such holistic perspectives of the wider organisational world were too general to help individual organisations or projects and so there was a need for understanding processes. Operational research provides this by bringing rational and numerical modelling techniques to issues of production. As this was focused on client needs and on a desire to design improved systems, it is possible to see this also as an early attempt at business process re-engineering. The process mapping helped them to detail the interdependence between the organisations and role holders, and this is a forerunner to supply chain management starting not in logistics but more concerned with effective delivery of services. The relationship between machines, people and organisations was not well understood and it was this that the Tavistock was making its mission. It continued to develop socio-technical systems[14] which became its marketable service as well as its theoretical position.

The open system analysis of the Tavistock meant that the BICRP was an abstract study but at an operational level. At this level, context is important and abstraction has to deal with this. In addition, relationships between individuals are also evident and so the natural conflicts of life are exposed. What systems analysis does is to focus on wider interactions in organisations yet attempts to relate parts of operation and context to the whole organisational performance. Thus, problems are not viewed as simple cause and effect results, as is commonly the case in other reports, but as systemic issues, that is, as a product of the whole structure and the way this structure is related. In this perspective, it is through communications that parts of the structure of an organisation relate to the whole both as regards purpose and as regards product. This was not understood then, and poorly understood now, as the majority of studies assume an intervention into a static world rather than into a dynamic business environment. They determined that the particular dynamics of the industry required the structure and relationships that were evident, however apparently malfunctional. Thus, the complex interdependence that was found with roles and relationships that were unstable and confused was a necessary part of the whole system. This could only be coped with through informal rather than through formal communications as many of the value conflicts are created by the system and so are legitimate. The Tavistock researchers' solution, that the industry needed to work together, was not new, and it was partly achieved during the BICRP itself. It is only in the last 6 years that there have been a number of collaborative bodies established. In particular, there has been a desire to stimulate networks, some again researched by the Tavistock, and also the M4I agenda where their use of showcasing demonstration projects celebrates together the success of the industry.

6.8   Conclusion

For 3 years the Tavistock Institute and the Institute of Operational Research had worked within a novel collaborative network in what had been previously a strongly dissociative setting. They succeeded in developing a very much stronger theoretical understanding of the operation of the industry at project level through the tool of socio-technical systems. This balance between rational analyses and the social psychology of complex organisations such as construction was unique then and continues to be valid today, and it can be seen in the current notions of business process re-engineering and supply chain management. The analysis became the focus of a wider reassessment of operational research[15] which established new ways of working in complex, uncertain and conflict prone situations. The OR tools were not designed for this world of fuzziness and change. They certainly did not deliver solutions as had been promised, and therefore the whole project started to breakdown. This lack of practical outcome along with the project relationship problems and industry economics led to conflict that consumed much energy. The industry did not change as a result of the project, yet the project raised issues of collaboration, learning and a process view of construction, returned to much later by Latham[16] and Egan[17]. Maybe such radical thinking takes 30 years to become the next generation's innovation.

6.9   References

1   Woodbine Parish, D. E. (1964) *Building Research and Information*. HMSO, London.
2   Banwell, Sir Harold (1964) *The Placing and Management of Contracts for Building and Civil Engineering Work*. HMSO, London.
3   BICRP (1966) Archives of the BICRP, at RIBA in London.
4   Marwick, A. (1995) *The Penguin Social History of Britain: British Society Since 1945*. Penguin Books, London.

5   Childs, D. (1992) *Britain Since 1945*. 3rd edn. Routledge, London.
6   Hillebrandt, P. (1984) *Analysis of the British Construction Industry*. Macmillan, London.
7   Ball, M. (1988) *Rebuilding Construction*. Routledge, London.
8   Emmerson, Sir Harold (1962) *Survey of Problems before the Construction Industries*. A report prepared for the Minister of Works. HMSO, London.
9   Stringer, J. (1967) Operational research for multi-organisations. *Operational Research Quarterly*, 18(2), 105–20.
10 Bryant, D. T., Mackenzie, M. R. & Amos, W. (1969) The role of the client in building. Document no. IOR/355/2. The Tavistock Institute, London.
11 Cherns, A. B. & Bryant, D. (1986) Studying the client's role in construction. *Construction Management and Economics*, 4.
12 Friend, J. & Hickling, A. (1987) *Planning Under Pressure: The Strategic Choice Approach*. Pergamon, Oxford.
13 Von Bertalanffy, L. (1968) *General Systems Theory*. Braziller, New York.
14 Emery, F. & Trist, E. L. (1960) Socio-technical systems. In: *Management Science, Models and Techniques* (eds C. W. Churchman & M. Verhulst), Vol. 2, pp. 83–97. Pergamon.
15 Rosenhead, J. (ed.) (1992) *Rational Analysis in a Problematic World*. Wiley.
16 Latham, Sir Michael (1994) *Constructing the Team*. HMSO, London.
17 Egan, Sir John (1998) *Rethinking Construction*. Department of the Environment, Transport and the Regions, London.

Chapter 7
Large Industrial Sites Report (1970)

David Langford

7.1 Background

1970 was a momentous year. The world was riven with conflict. In Vietnam the American bombing campaign led to continued student protests throughout the world. It was to be five more years before the war was ended. In Ireland the British troops that had been sent in 1968 as a protective force for Catholics were now engaged in the middle of a bitter civil war. In the Middle East, Israel and Egypt continued hostilities. In South Africa the apartheid regime was almost universally disposed, and anti-apartheid movements had a worldwide presence. The Womens' Movement was emerging, with feminism proclaimed as a powerful political and social agent of change. Black consciousness was stirring with the authority of Christian-based movements being displaced by a more militant creed.

In popular culture the major story was the break-up of the Beatles. Helpline phones were not set up. Top of the Pops heralded a wide range of songs from the lugubrious Lee Marvin with *Wand'rin' Star* to the effervescent and durable Mungo Jerry's *In the Summertime*. At the cinema, Oscar nominations were shared by the cerebral *Five Easy Pieces* to the more maudlin *Love Story*. The winner was *Patten*. It was also a momentous year for sport. England went to Mexico to defend the World Cup, only to fall 3–2 to Germany in the quarter finals. Peter Bonnetti in goal was held responsible by the tabloids. Chelsea had won the English First Division and Aberdeen held sway in Scotland with Cup and League titles. The mercurial France held the Five Nations Rugby trophy. In cricket, the South Africans were due to tour, but the Government pressurised the cricket authorities to withdraw the invitation; Kent had won the County Championship. The more fashionable would have worn flared trousers or miniskirts.

The Government, under Prime Minister Harold Wilson, wavered about when to call the general election: Spring or October. Crossman[1] records that a key issue in deciding the date was the World Cup. By delaying until October it was hoped that an English victory would be as beneficial as it had been in 1966.

The Government of the day had several concerns: inflation was running at 20% and workers and managers were cajoled to increase productivity, but the dominant debate was about how industrial relations were to be managed. In the 2 years preceding 1970, the Government, unions and employees had engaged in acrimonious exchanges about the reform of industrial relations. Barbara Castle had launched a White Paper, 'In Place of Strife', which sought to provide a much more controlled environment for the prosecution of industrial relations. Eventually, the trade unions pushed back the intended legislation, which had to wait until the Conservatives introduced the Industrial Relations Act as one of their signature pieces of legislation in the early days of the 1970–74 Government.

Within this climate, the engineering construction industry was beset by industrial relation problems. The Government, concerned about the impact of delays and cost overruns on large engineering construction projects, set up a working party on large industrial construction sites under the auspices of the National Economic Development Council (NEDC). It first met in July 1968 and reported in May 1970. The report was entitled *Large Industrial Sites*. The terms of reference for the Committee were to[2]

'enquire into the problems of organisations of large industrial construction sites with particular reference to labour relations to investigate their causes and their effects on the cost of commissioning and operating plants, and to make recommendations.'

7.2 The context of the construction industry

7.2.1 The workforce

The numbers employed in the construction industry stayed at over 1 million workers throughout most of the 1960s but 1969 was the cusp in employment trends. Recession bit, and so in 1969 numbers fell below 1 million and continued falling through 1970 and 1971. The workforce employed on new industrial construction followed a similar trend. Workers in new industrial construction numbered around 0.5 million until 1968, by which time they began to decline. It must be noted that not all of these 0.5 million workers would have been employed on large industrial sites. The electrical and mechanical trades employed around 50 000 workers. What is noticeable is that the recession of 1968 ate into the labour force at a rate of 6–8% reduction in workers per annum. In new industrial work, attrition in 1968–70 occurred at a rate of 3% until 1970–71, when it shot up to 10%. How may this be explained? Industrial work is likely to constitute major undertakings of long duration and so act to bridge a recession. Figure 7.1 illustrates these statistics. In addition,

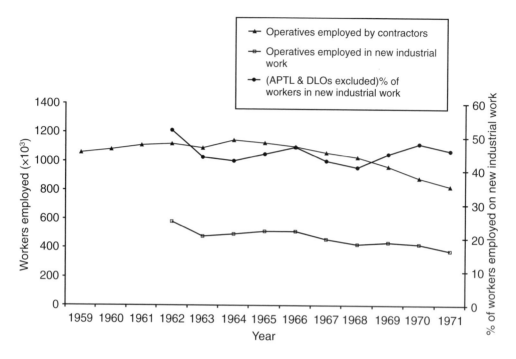

Fig. 7.1 Number of construction workers employed 1959–71 (APTL: administrative, professional, technical and clerical; DLOs: direct labour organisations).

much of the industry was funded by Government which could be selective in releasing projects so as to smooth out demand in the industry

The recession was reflected in the unemployment figures for building workers. Unemployment rose sharply between 1966 and 1967 and crept up to almost 100 000. As regards the trades associated with industrial construction, they constituted some 11–13% of the unemployed. Again, these data highlight the resilient nature of industrial construction trades. They made up 40–50% of the total workforce but experienced only a quarter of the unemployment figure.

Vacancies for construction craftsmen over 18 shrank from 12 000 in 1966 to 4000 in 1970. Growth in wages slowed in the face of the recession. Using 1965 as an index of 100, the ratio of earnings for all industry and the construction sector shows the impact on living standards. Through the period 1960–65 the index of the earnings of all manual workers showed that construction lagged behind all workers by 1–3%. From 1966–9 this feature was reversed, with construction leading by 1–2%. In 1970, construction had fallen behind badly and trailed the all workers index by 8%.

Moreover, in the period 1960–70 construction workers, when compared with all industry averages, had to work an extra 2 hours per week to generate these earnings.

7.2.2   *Output*

In 1966 the Government enacted the Industrial Development Act. The purpose of this legislation was to promote investment in goods which would increase industrial output.

The Act allowed tax breaks on investment in capital goods, and naturally industrial construction benefited from this advantageous tax regime. The whole industry was based around a high proportion of construction expenditure being driven by public sector spending. Much of the Government investment was dedicated to industrial construction in publicly held utilities (water, gas and electricity plants) or industries that were nationalised in full or in part (British Steel and British Gas). Naturally, the Government had a strong interest in having a productive and harmonious engineering construction industry. Output in the whole of the construction sector grew from over £4 billion in 1966 to almost £5.5 billion in 1971. The output of new industrial construction was some 10% of total activity. Of equal interest are the indices of cost and production. These show that costs were rising at a faster rate than productivity. This, then, is the background and context of the report.

7.3   The report

The Committee was large by comparison with today's task forces. It had 36 members drawn from the great and the good of Government, trade unions, employers' associations and senior managers drawn from the engineering construction industry. (By comparison, the Egan report looks lean with only 10 members and only one trade unionist participating.) The Chairman was Angus Paton, President of the Institute of Civil Engineers; Lewis Wright, formerly the Chair of the General Council of the Trade Union Congress (TUC), was the Deputy Chair. The Committee started life as a rather smaller group but quickly expanded to afford representation to the diverse groups that populated the engineering construction industry. From the Union side the big battalions of the Amalgamated Union of Engineering Workers (AUEW) and the Transport and General Workers Union (TGWU) dominated, but smaller groups such as the Electricians and Plumbers Union and the Sheet Metal Workers Union also saw participation. From the employers' side, the Electrical Contractors Association, the powerful Engineering Employers Federation and the Heating and Ventilating Contractors were present. A numerically small but politically strong client group, the Oil and Chemical Plant Constructors Association, led the debate about the changes they would like to see for the industry. With regard to the mainstream construction employers, the Federation of Civil Engineering Contractors and the National Federation of Building Trade Employers were both invited to participate but declined, arguing that industrial relations (the main prompt for the report) were slightly different outside of the engineering construction sector.

The Committee recognised early on that a group of 36 was unwieldy and so broke itself up into two subgroups. One, under the direction of Paton, looked at the 'problems of management by client and contractor'. The question of 'labour problems' was addressed by a group chaired by Wright. These groups were still large and membership overlapped.

What had brought the Working Party together? A confluence of events had prompted action. Primarily the problems in bringing power stations in on time and budget had been recognised as early as 1966. The Ministry of Power (then a Government Ministry not a nightclub!) had considered the problems in its commissioning of Central Electricity Generating Board power stations. At the time, Government, concerned with national planning, had established the National Economic Development Office (NEDO) which delegated industrial planning to sector-specific committees, i.e. the National Economic Development Committee for Construction. The Economic Development Committees for Electrical Engineering and Chemicals both flagged the issue of delays as meriting special study.

The first task of the Committee was to define what constituted a 'large industrial site'. The Committee found it difficult to assign an all-purpose meaning to the phrase. Consequently, it saw the industry as a three-headed hydra, two of the heads being characterised by client type and the third by size of site. The first coverage is shaped by 'mechanical and electrical construction', which was used to define an industry in which much of the work took place in factories and construction sites and constituted two sets of workforces, those factory based and those site based, each doing similar work but under very different terms of employment. The second definition, construction in the process industries, saw the industry from the level of the site. The Committee discovered, through its specialist statistical group, that the value of site work on large mechanical and electrical projects was about one-third of the total investment in industrial construction and added 10% of the UK's annual fixed investment. The third framework for the industry was to consider the industry defined by four characteristics, namely:

- Size – projects in the range of £50–100 million.
- Skills – the range of crafts used are similar: erectors/riggers, pipefitters/plumbers, electricians, fitters/platers and boilermakers, not to mention the contingent of civil engineering crafts.
- Structure – the sites had many contractors which had direct relationships with the clients. The size of the contracts varied enormously.
- Supervision – managers needed to build a fresh workforce for each project and coordinate the work of many disparate constructors on projects where the technology being built was cutting edge.

7.3.1 The nature of the problem

In a study of clients, the Working Party found that:

- In a study of 13 power stations, cost overruns ranged from –22% (presumably because of changes in scope) to +50%. The average cost overrun for projects that showed escalations was 22.7%.
- Delays on the power stations ranged between 5% and 40% with an average delay of 20%! That is to say, a typical project of 70 months duration would be delayed by 43 months.
- In a study of cost overruns in 13 oil and chemical plants the range was 12–33%, with an average of 15% (four of the 15 projects were brought in under budget).
- Delays to oil and chemical plants ranged from 4 to 58%. The average was 26%. Four of the projects were completed on time.
- In oil gasification plants constructed in the period 1961–70, for a sample of seven projects the cost overruns ranged from 0 to 23%, with an average of 11%. Delays ranged from 0 to 23 months. The estimated duration is not given.

Outside the construction of power stations, most of the clients for large industrial projects are multinational and/or chemical companies. Such organisations are fickle in selecting their locations for such plants. Certainly, the tax breaks offered by the Industrial Development Act encouraged companies to locate in Britain, but the cost overruns and delays experienced and moreover the uncertainties in delivery cost and time made investment decisions problematic for the UK economy. The fear of relocation was ever present.

7.3.2 Why did these overruns occur?

Certainly, the characteristics of engineering construction projects lend themselves to these problems. Firstly, capital cost estimates are often deflated by the project champions in order to promote the project. Connaughton[3] notes that cost estimates at the early phase are largely political guises. The estimate has to be accepted by the client's capital expenditure committee or another group that will sanction the go-ahead. The estimate must be close enough to the out-turn costs not to embarrass the champion or the sanctioning group but is anticipated to be above the first estimate. More mundane reasons will also be self-evident; complexity is always a challenge, and subassemblies requiring manufacturing to take place nationally or internationally will mean that this kind of project is hostage to knock-on increases in costs and possible delays. The long timescale for the projects, which are technically challenging and frequently subject to redesign through breakthroughs in production or process engineering, and vicissitudes of market prices for the commodities being produced are further factors. With turbulent industrial relations also taken into account, the performance of such projects is always likely to be problematic. In a survey of contractors who were asked to rank the reasons for delay on such projects, they reported 'late delivery of materials or plant' as being the most common factor, marginally ahead of 'late design changes'. 'Labour disputes', the factor that prompted the report, scored relatively modestly. The full list is shown in Table 7.1.

Table 7.1 Cause of delay in construction of large sites.

Reason for delay	Total score[a]	% of companies ranking as first
Late delivery of materials or plant	180	21
Late design changes	173	27
Unexpected low labour productivity	120	10
Labour disputes	88	8
Delays in subcontractors	79	8
Skilled labour shortages	61	5
Faulty materials	24	1
Faulty workmanship	12	—
Other reasons	89	17
Other contractors' performance	35	8
Access	28	7
Client's performance	14	1

a The four reasons were ranked by contractors in order of importance, and were subsequently scored 4 3 2 1.

7.4   The content of the report

The final report is broken down into five chapters. Chapter 1 considers the role of the client and contractor in managing projects. Chapter 2 looks at the training of managers and operatives, and Chapter 3 considers the time management of large industrial sites. The heart of the report, about industrial relations, is found in Chapter 4, and the report concludes with recommendations for action in Chapter 5.

7.4.1   *Chapter 1 – Management by client and contractor*

The Committee recognises that the delays and cost overruns in part stem from decisions taken by the client before work on site begins. Critical choices for the client are:

- the form of contract;
- the number of work packages to be let;
- the extent of the client's input into the design of the plant;
- the amount of allowable variations to the contract and how any variations are to be controlled;
- the liquidated and ascertained damages applicable or any countervailing incentives for the contractor;
- how the project is to be supervised by the client.

The report sees these questions as a set of interlocking issues where decisions in one area have impacts and implications in other areas. The selection of work packages is taken as an example. Breaking down the project into many small specialised packages may have advantages in that specialisation may be reflected in good performance, but the greater the number of packages, the greater is the requirement for sophisticated management to coordinate a complex set of interactions on and off site. In construction, problems always occur at the joins, be they between the wall and roof or between subcontractor packages.

By breaking down the project into small units, the project is likely to face labour difficulties as a result of different contractors employing the same trade with different pay and conditions, thus exacerbating latent industrial relation problems. In response to the questions posed at the start of this section, the Committee recognise that there is unlikely to be an 'industry recipe' by which all projects are managed but that sets of compatible policies across all of the issues provide an opportunity to optimise the effectiveness of project management. In many ways the advice given follows the contingency management theories of Lawrence & Lorsch[4] who see the external environment as a key issue in setting managerial regimes. If it were hostile, then greater integration of the management functions would be necessary, whereas if it were more benign, then greater differentiation could be tolerated.

The report rehearses the arguments for different forms of contract and presents three alternatives: lump sum contracts, reimbursable contracts and selective tendering and negotiation. It does not offer a dirigistic model for selecting contract types, rather it urges clients to be more minded to consider reimbursable and negotiated contracts.

The conclusions are prescient. It was argued that projects funded by the public sector should be free to negotiate arrangements between clients and contractors and that rigorous prequalification procedures should be in place. In terms of project strategy, the report suggests that a management contract type of arrangement is the most suitable, given that clients are better able to manage a smaller number of contractual arrangements. Accompanying the measure is the suggestion that clients would be better served by firms that had greater integration of manufacturers and site-based installers. This call is still resonant of the industry today, where clients often call for greater consolidation of the construction industry with its opportunities for greater integration of the process.

The vexed issue of variations is raised. As was seen earlier, the survey of contractors indicated that variations and design changes create delays. The report's recommendation is that concurrent engineering should not be used, as overlapping of design and construction amplifies the problems of variations. It suggests that a tight size gate system be put in place by clients so that only variations that are absolutely necessary are approved. A hierarchical system of approving variations is suggested, with more senior management being involved in approving higher-value variations.

The report is very dismissive of the benefits of liquidated damages (called 'penalties' in the main document) or incentives to contractors for early completion or effective cost management. It is more persuaded that incentives have greater latent power to promote effective performance than the fear of penalties. This finding is at odds with the preference for the use of reimbursement contracts where a strong feature is a sliding scale of payments for the contractor which rewards achievements in delivering good performance. Where the report is more certain is in the role of the client or managing contractor in providing common services such as cranage, lifts, hoarding, canteens and welfare facilities and other amenities that will be used by all participants in what is invariably a multicontractor environment. The provision and maintenance of such facilities will be the responsibility of the client's project management team. In its recommendations, the report builds on this point by being prescriptive in writing up a welfare code offering a 'high and consistent standard of amenities on all sites'. This code suggests that these facilities should include medical staff, shelters and drying rooms, washrooms, protective clothing, canteens, transport to and from the site, car parking facilities and boot strap camps. The role of the client's project manager is to stimulate a discussion of the most effective organisation structure to deliver those benefits. Here, the kernel of a debate within the Committee can be observed. Most members recommend a project structure with the site-based managers being given considerable authority over technical issues. Evidently, a minority view is that the functional specialists hold sway over their area of technical expertise. The balance between rapid solutions that do not hold up the job and technically 'correct' solutions provided by the functional specialists exercises the Committee.

7.4.2   Chapter 2 – Training

Chapter 2 focuses on the training of managers and operatives for the engineering construction industry. In 1970 the Industrial Training Act 1964 was in place, and this Act had realised the concept of industry training boards which were funded by a levy on the payroll of companies whose turnover determined the extent of the levy they had to pay. The training function for the engineering construction industry was covered by two training boards – the Construction Industry Training Board (CITB) and the Engineering Industry Training Board (EITB). The report finds this division of responsibility deleterious to the training strategy of the industry. It firmly opts for the responsibility for training to be given to the CITB. The report sees a necessity to train managers and operatives.

Management training

The report bemoans the absence of bespoke training courses for managers responsible for delivering mechanical and electrical construction. In passing, it revives the idea a 'staff college' for the training of construction managers, first mooted in the Simon report of 1944[5]. The provision of such training, in the eyes of the Committee, would elevate the status of managers in the industry and with it the salaries paid to managers. They cite that managers frequently earn less than the workers they are supervising. (On a personal note I worked as manager on an industrial site – a cement works – and was trying to develop a carpenter into a foreman. He thanked me for my attention but indicated firmly that he had no aspirations of being a foreman. When asked 'why', he noted that, while he would leave the job at 5.30 PM, I was there until 6.30–7.00 PM for no extra pay and would probably be worrying about the job for longer and our wages were broadly similar. The extra responsibility of being foreman was not matched by additional compensation.) Cross-company training is recommended in order to bring clients and contractor staff together for courses at selected 'partnered' educational institutions.

In the intervening 30 years since the report, progress on management training has been slow. An inspection of university prospectuses reveals opportunities in Golf Course Management, the University of Florida evidently runs a BSc in Surfing, but no specific university courses on management in the mechanical and electrical contracting industry could be found.

Operative training

In an industry said to be riven with industrial disharmony, one issue unites the unions and employees: the need for the industry to provide systematic training for operatives. It is notable that the craft labour needs of the industry are provided by staff from other industries. As such, the workforce arrives at site with only a smattering of the knowledge required to make a good engineering construction worker. At the time, the Government recognised that the shortages of skilled labour could hamper economic growth, and so it set about converting unskilled workers into craftsmen by providing courses of 6–12 months duration at the newly formed Government training centres. Not only was training provision in craft skills in short supply, but training in support of an improved industrial relations climate was urgently required. The beneficiaries of this activity were firstly to be line supervisors and shop stewards and then the bulk of the labour force. The reasons for this emphasis are both obvious and machiavellian. At the level of the obvious, the notoriety of the poor industrial relations achieved on many large industrial sites demanded visible action. Underlying this need is the observation that industrial relations training would enable managers to gain swifter acceptance of new construction techniques and new ways of working.

It would be fair to say that progress has been made on the training of craftsmen in the industry, but the recent CITB survey of the needs of the industry point up the shortfalls in key trades making a contribution to the engineering construction industry.

7.4.3   Chapter 3 – Programming

The short Chapter 3 concerns itself with the programming of work. The report recognises that the classical hierarchy of plans is frequently in place, with master plans being drawn up by the client and the broad-brush work stages such as design, manufacture, construction and commissioning being used to formulate a framework plan. Such plans reveal the important milestones that are incorporated into tender documents or information used in negotiating contracts. The recommendations urge that contractors submit more detailed programmes with their tenders; this suggestion was precocious for it was not until the 1980s that the Institute of Civil Engineers Contract and the Joint Contracts Tribunal standard forms required tenderers to provide a network-based programme to be submitted as part of a contractor's tender. The report notes that a budget of 0.5% of the total value of the project should be set aside for project planning. The form that this planning should take is also debated, and network methods are recognised as providing a superior instrument for planning and coordinating multicontractor environments. However, it is recommended that the good practice of translating these networks into short-term plans (weekly or daily) that are more accessible to craftsman on site be paid greater attention.

Successful planning depends on five characteristics, namely:

- Senior management in client and contractor organisations need to be aware of planning techniques and committed to using them.
- Effective communication of planning information between client, contractors and subcontractors.
- Rapid updating of networks through ready access to computer processing facilities.
- The presence of planning staff in the offices of clients and contractors.
- Training in the use of network planning needs to be extended.

The old shibboleth of differentiated functions of design, manufacture and construction adversely influencing planning is cited as a reason for the absence of integrated planning.

In all, the section on programming is bland and adds little to the report.

7.4.4   Chapter 4 – Industrial relations problems

The heart of the report deals with industrial relations on site. As Morris[6] points out, projects at this time, especially ones driven by energy production, were driven by the need for early delivery and cost pressures were secondary. Oil-based projects required early delivery in order to profit from buoyant commodity prices. In the public sector, more emphasis was placed on the tender price, always elicited by selective competition, rather than the out-turn costs which were frequently much greater than the tender price.

  This need for quick project turnaround influenced the balance of power between labour and management. At the time, industrial relations were conducted in a hostile way. It was only 2 years since the Paris spring of 1968, when workers fuelled by syndicalist sentiments had occupied factories, much to the chagrin of the French Communist Party. The events in Prague inspired a sense of what could be possible if left alone to self-determine this practice. Interest in workers' control, workers' cooperatives, autonomous work groups and worker representation was emerging strongly among a well-organised working class. The influential Royal Commission on Trade Unions and Employers' Associations (the Donovan report[7]) had also reported 2 years previously, and this had provided the backdrop for the *Large Industrial Sites* report. At the time there was a massive growth of workplace, as opposed to national, bargaining; the growth of work-based trade union representatives endorsed this trend. Topham[8] records that in 1968 there were some 175 000 shop stewards, yet by 1971 this had doubled to 350 000. This growth reflected the authority of the trade unions and the autonomy of shop stewards to bargain on behalf of the trade unions at site level. In short, a spirit of confidence and militancy was in the air. This mood was influencing the willingness of workers to press their demands through strike action. Although construction and manufacturing were only considered as 'medium risks' in terms of a scale of the propensity to strike[8] (the scale had five bands from high to low propensity to strike), the reality was that the medium category contained industry, which showed a greater variability in attitudes to industrial action. Workers in engineering construction may be said to have been organised and militant. Most of the mechanical and steelworking trades were organised by the Amalgamated Union of Engineers which had merged with the Construction Engineering Union in 1967. This was a militant union. The record of days lost to strikes in engineering construction in 1965 was 911 per 1000 workers. The construction industry as a whole was much more sanguine, and only 150 days per 1000 workers were lost. Only the motor industry, shipbuilding and road passenger transport lost more days. The long-term trend for stoppages in construction shows an increasingly militant workforce. Figure 7.2 shows the days lost in the period 1960–71.

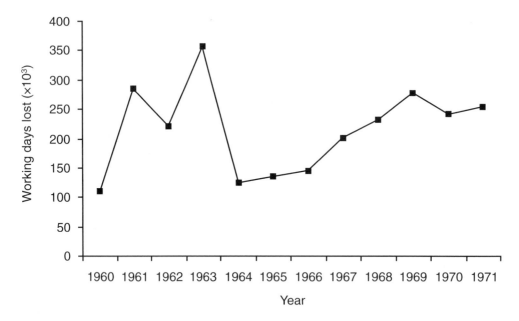

Fig. 7.2   Working days lost to strikes in construction in the period from 1960–71.

The causes of the stoppages in industrial construction were variable. Naturally wages issues dominated. The stoppages were as follows: wages (bonus) 24%, other 17%, dismissal 19%, site conditions 8%, working practices 7%, sympathy 7%, demarcation 5%, redundancy 3% and other matters 12%.

What was the Establishment to do? The Labour Government's response was to launch a White Paper, 'In Place of Strife', which sought to regulate industrial relations. The White Paper polarised Cabinet and Party. The Left in the Cabinet vehemently opposed the proposed reforms, and the trade unions, with the support of the future Labour Prime Minister Jim Callaghan, were extremely hostile to the move. Eventually, the White Paper was withdrawn, and it was not until the election of the Conservative Government in June 1970 that the legislative control of industrial relations became centre stage through the Industrial Relations Act (1972).

The industrial construction industry, like most others, was faced with an environment that encouraged 'haphazard local bargaining (where) earnings vary without equity or logic within and between sites'[2]. The evidence for this variability of earnings is stark. Average earnings, excluding allowances for trades employed by different contractors on one site, varied by 70% (erectors/riggers) and 67% (welders). Disparities of earnings occurred between sites and within sites. Such anomalies were clearly going to be a source of friction. The heavy dependence of bonus payments to make up a wage exacerbated the situation. The bonus in power station construction was typically 20–30% of the take-home pay. When such a large proportion of the pay is incentivised, then disputes are inevitable. The oil and petrochemical industry, driven by its powerful employers' association, the Oil and Chemical Plant Construction Association, sought to regulate this by using site agreements to 'produce more uniform and firmly applied systems of pay and conditions for the sites to which they relate'[2]. Thus, the industry was beset with two approaches to managing industrial relations: the private sector was covered by site agreements that sought to unify wages and conditions across employers and trades; the public sector in power and steel plants relied upon various national agreements in which a lower basic pay was supplemented by bonus, overtime and other emoluments. The position in the public sector was exacerbated by the variety of national agreements that covered various trades operating in the industry. Electricians, erectors, heating and ventilating engineers and civil engineering trades were all covered by separate national agreements.

What was to be the solution to this problem? For some time the employers had been agitating for a single agreement covering a narrow range of issues. The report provided comfort for this view. The proposed national agreement would have symbolic as well as practical consequences. Any agreement would acknowledge the presence of an industry that was distinct from construction and manufacturing, and consequently a separate agreement would provide a centrifuge for employers and trades unions. The forum created by the agreement would enable the industry to harmonise wage rates, employment conditions and benefits and so curtail the practice of 'leap-frogging' by which different groups of workers made different claims at different times for different concessions. It was believed that the benefits of this homogenisation of conditions would improve industrial relations by reducing stoppages and encouraging allegiance to an industry and so discourage casual employment. The political fall-out from such a proposal was stark. The Trotskyite 'rank and file' movement of workers and shop stewards, while numerically weak, was influential in the industry, and they saw this as a corporatist alliance between trade unions and the employers. The Amalgamated Union of Engineering Workers saw benefits in that the officials of the union – usually drawn from the Left, and strongly influenced by the Communist Party – would be able to exercise greater authority and control over the workforce. The clients were also seen as beneficiaries. A national agreement would provide greater cost certainties because contractors would be better able to predict labour costs. The consensus for a national agreement was found after considering alternatives to formulating 'site agreements'. Employers, unions and clients were wary. For the employers, site agreements signalled the prospects of labour cost escalation but, more importantly, the turbulence preceding an agreement at the start of each site would be difficult to deal with. A hostile start to a job could pervade the atmosphere for the duration of the project. Thus, while local, site-based bargaining was acknowledged to be important, it was seen as subordinate to a national agreement.

In particular, site-based bargaining was seen as important in controlling bonus systems and overtime working. The panel were apparently very keen to promote the retention of bonus schemes, and there is evidence that it saw benefits in harmonising bonus schemes such that there were pan-industry uniform standards for bonus rates. In such a world, workers performing at a certain level would earn similar bonuses irrespective of the site on which they worked. The practical difficulties in setting time standards for elements of work were acknowledged as a major hurdle, especially when work study was not a deeply embedded practice in the industry.

However, the employers see incentivised pay as a protection against slacking by the workforce. The issues deemed appropriate for national agreements and those suited to local bargaining are shown in Table 7.2.

Table 7.2 The National Agreement – matters for determination.

At national level	At site level
Basic rates and differential pattern for skills	Bonus schemes
Hours of work	Productivity payments
Premiums for overtime and shiftwork	The removal of demarcation practices
Holidays and holiday payment	Redundancy
Travelling and subsistence	Special payments of allowances for unique site conditions
Recognition and facilities for stewards	Detailed steward or union facilities
Settlement of grievance affecting 'national' policies	Domestic grievance procedure
Code of amenities	Control of overtime and the use of shiftwork
Code of dismissals, safety, etc.	

7.4.5   Chapter 5 – Proposals for follow-up

If such a national agreement were to be established, then it would need an agency to oversee its implementation and to police its operation. This putative National Joint Council (NJC) for Engineering Construction would, in addition to its normal negotiating functions, concern itself with training, grading and status of employees. The progress in establishing such a structure is dismissed in the next chapter.

The report sets out an agenda for the NJC to address. This is as follows:

Travelling and lodging arrangements

The report urges that negotiations in the national agreement unify travel and lodging arrangements in order to eliminate the 'under the counter payments' prevalent on large industrial sites.

Demarcation

> This is not recognised as a major problem but merely as an irritant. The report urges the trade unions to merge and eliminate 'who does what' disputes. The panel recommends that dispute procedures be incorporated into a national agreement to define mechanisms for resolving demarcation disputes.

The role of shop stewards

> The saloon bar view of shop stewards of the time was that they were fomenters of trouble. They were frequently demonised by the press. 'Red Robbo' ruled the roost at the Longbridge car plant in Birmingham. A shop steward at Heathrow was branded as 'the most dangerous man in Britain'. (On a personal note, this man used to drink in my local pub. As he stood at the bar he looked avuncular. Always accompanied by his tartan-coated little Scottie dog on a lead, he presented a sharp contrast to the picture painted in the press.)
>
> Against this backdrop, the report presents a refreshingly different idea of a shop steward. It calls for the proposed National Agreement to document the role and remit of shop stewards and the facilities to be provided. It urges training facilities for stewards to be provided by the unions and the employer.

Grievance procedures

> The turbulent environment on large industrial sites led to many of the grievances that resulted in poor performance. The report recommends that a grievance procedure be established which would apply to all sites. As incentive schemes were the source of many of the disputes, it was argued that unions and employers should have bonus clerks to agree targets for incentive payments.
>
> The report is prescient in seeking to engender an industrial environment where workers could enjoy greater security of employment. It rejects the traditional employer's view that casualised labour produces a hungry and more docile worker by acknowledging that

> > 'hire and fire policies foment hostility, prevent the formation of efficient work teams, render training uneconomic, make safety education more difficult and block the introduction of more enlightened personnel policies.'

The call was for the industry to decasualise by contractors increasing the size of the permanent workforce. At the time, only 37% of craftsmen employed in the industry were regarded as permanent employers and only 18% of the workforce had experienced over 5 years service with their present employer.

The report calls for the establishment of registered workers for the industry. This plan formed part of the Labour Government's agenda until it was dropped from the legislative programme in 1969. Fellows *et al.*[9] observe that the Construction Industry Manpower Board and trade unions urged compulsory registration but that employers resisted this move.

7.5   The impact of the report

The report concluded that a strong committee made up of various representatives of the industry should be established. One of the tasks with which this
steering group was charged was to launch an NJC for the industry. One of the
first acts of the steering group was to popularise the contents of the report
and launch the pamphlet *What's Wrong on Site* in July 1970. This was expected
to galvanise employers, unions and the workforce to press for the *Large Industrial Sites* agenda[2]. Early efforts were directed towards training, and 1 year
following the launch of the report NEDO[10] had stimulated an initiative to
improve training in industrial relations. Companies were encouraged by the
training boards (CITB and EITB) to put on courses in industrial relations and
to ensure that these catered for site-based managers. In many instances this
happened. However, it was not until 1982 that a national working rule agreement, similar to the one enjoyed by the building industry, was available to the
engineering construction industry. The principal signatories to this agreement were the Engineering Employers Federation and the AUEW.

Fellows *et al.*[9] saw the agreement as providing the framework for the ambitions
of the *Large Industrial Sites* report[2]. It was expected that the agreement would

> 'give greater consistency of wage rates, earnings, site conditions and bene-
> fits, thus reducing chain reactions of claims and disputes. A more ordered
> industrial relations climate could result in workers and supervisors recog-
> nising the requirements of the Large Site Agreement and, hence, move-
> ment between sites could be more ordered, with a consequent
> decasualising of the labour force. Some have argued that the strengthened
> agreement for large sites would enhance the authority of the official trade
> union leadership at the expense of site-based shop stewards.'

Paradoxically, this national agreement had anticipated that it would be
supplemented and indeed supplanted by site-based agreements. Such site
agreements could override the national agreements binding contractors and
subcontractors and so override inconsistent agreements operating in a
multiemployer site. The advantages of site agreements were cited in Fellows
et al.[9] as:

(a) negotiation at the start of a job can be anticipated in order to reduce
    questions of demarcation and productivity;
(b) overtime can be controlled;
(c) the amount of shiftwork can be determined;
(d) the facilities to be available for shop stewards can be decided;
(e) a policy can be formulated on the selection for redundancy.

What was the reaction of the trade press and academic press to the report?
The evidential trail is rather cold, but inspection of *The Engineer*, a magazine
billing itself as the magazine for engineering managers, did not cover the

story. Likewise *New Civil Engineer* (formerly *Civil Engineering and Public Works Review*)[11] carried one pithy paragraph which noted:

> 'new policies in mechanical and electrical plant construction could result in benefits running into many millions of pounds. A guide to these new policies has been put forward in the NEDO report on large industrial sites.'

The *Building* magazine of the same year was as chary[12]. Again in a one-paragraph report, it records that the delays experienced in large projects were not the result of poor industrial relations but that management policies were the primary problem. It observed that clients needed to be more insistent on unified policies among all contractors on site.

Among other professional magazines, *Chartered Surveyor* remained silent and *Engineering Magazine* had nothing to say. The impact on the academic press was also minimal. While the *Industrial Relations Journal* and the *British Journal for Industrial Relations* carried papers on one of the report's major findings, the training of shop stewards[13, 14], no specific mention is made of the *Large Industrial Sites* report, nor for the that matter does any paper feature the construction industry. For managers, the British Institute of Management[15] reported on the training of managers for industrial relations duties.

By October 1970 the political background for industrial relations had changed. The Conservatives were elected and their Industrial Relations Act of 1972 had polarised employers and the trade unions. In 1972 the first national building workers' strike was called over pay, and this involved engineering construction. The lasting memorial to the report was the large site agreement which is still in use today. The ambitions of a less casualised workforce had to wait until the late 1990s before the Inland Revenue were pushed into action to encourage a greater use of directly employed labour. The report's ambition of a 75% permanently employed workforce still remains a distant aim. The welfare code that the report pioneered was ahead of its time, and the 1999 Respect for People campaign may have said to have had its origins in the report.

The 1998 consolidation of the Welfare Regulations are largely modelled on the amenities that the report envisaged as being standard facilities on each site. The issue of registration of workers is still outstanding, but movement in this direction has been observed.

Thus, to sum up, the report heralded changes in the industrial relations climate and practice in industrial relations on large industrial sites. As a report it was farseeing and imbued trade unions and employers with a liberal approach to industrial relations. It is perhaps unfortunate that wider and stronger interventions into the industrial relations scene overshadowed more rapid progress towards its ambitions.

7.6   References

1   Crossman, R. (1979) *Socialism Now and Other Essays* (ed. D. Leonard). Cape, London; *The Crossman Diaries*. Methuen, London.
2   NEDC (1970) *Large Industrial Sites*. HMSO, London.
3   Connaughton, J. (1993) Investment divisions in large capital projects. PhD thesis, University of Greenwich, London.
4   Lawrence, P. R. & Lorsch, J.W. (1967) *Organising and Environment: Managing Differentiation and Integration*. Harvard University, Graduate School of Business Administration, Cambridge, Massachusetts.
5   Simon, Sir Ernest (1944) *The Placing and Management of Building Contracts*. HMSO, London.
6   Morris, P. (1994) *Management Projects*. Thomas Telford, London.
7   Donovan, L. J. (1968) *Royal Commission on Trade Unions and Employers Associations*. HMSO, London.
8   Topham, T. (1975) *The Organised Worker*. Arrow, Tiptree.
9   Fellows, R., Langford, D. A., Newcombe, R. & Urry, S. (2001) *Construction Management in Practice*. Blackwell Science, Oxford.
10 NEDO (1971) *The Training and Development of Field Managers in Engineering Construction*. HMSO, London.
11 *Civil Engineering and Public Works Review* (1970) July, 711.
12 *Building* (1970) News item. 2 May.
13 Young, D. & Findleter, J. (1972) Training and industry relations – a broader perspective. *Industrial Relations Journal*, Spring.
14 Withnall, A. (1972) Education and training for shop stewards: a reassessment. *British Journal of Industrial Relations*. Autumn.
15 BIM (1971) *Industrial Relations Training for Managers Working Party*. British Institute of Management, London.

Chapter 8

The Public Client and the Construction Industries: The Wood Report (1975)

Graham Ive

8.1 Overview of the period

Although published in 1975, the report's Working Party was formed in December 1971. Research data used in the study were collected from contracts awarded between 1969 and 1971. The gestation period of this report thus straddled not only a change in Government in 1974 but also, and far more important, the traumatic change from the final years of the post-war long boom or golden age to the period of political, economic and social crisis of the mid-1970s[1,2]. No sense of this collapse of established certainties can be found in the report, whose authors, not surprisingly, fail to see what only became clear in hindsight.

The report is a product of one of the key tripartite (state, large firms, trade unions) institutions that were central to the 'managed consensus' of the earlier period.

The 5 years prior to publication of the Wood report include the last years of Keynesian demand management by fine-tuning and witnessed a major example (in 1973) of the unplanned cuts in public capital expenditure that became a feature of the crisis period. These years represent a just-beginning long ebb tide from the high water mark in the share of public sector in total construction demand. This was still 45% of all new work orders (and 47% of total output) in 1973, while 1967 had been the peak for both volume of public sector orders and share of public sector in total orders. Demand was dominated by new construction rather than repair and maintenance (R&M) (contractors' total output, for public and private clients, in 1973 was £7100 million, of which only 22% was R&M).

Direct labour organisations (DLOs) were still an important source of public sector output (£830 million in total), especially of R&M[3]. Contractors' output of new public work in 1973 was £2492 million (£717 million housing and £1775 million other), compared with £3019 million of private new work. Contractors' output of R&M for the public sector in 1973 was estimated (Wood report) at £700 million (approximately £3200 million total, less £2492 million new work). The data given above are in current 1973 prices and are taken from Department of the Environment (DOE) statistics.[4]

Around 80 000 dwellings per year were still being demolished in Great Britain under slum clearance programmes. The public house building programme had just passed its post-war peak, though it was not widely realised at the time that never again in the twentieth century would public housing output approach its 1969 level. Much the same is true of many other public sector programmes.

It is also possible to look at the output of the early 1970s measured in terms of today's prices. For construction output at constant 1995 prices we have a choice of two series, one from the Office of National Statistics (ONS) and one from the Department of the Environment, Transport and the Regions/Department of the Environment (DETR/DOE). Expressed as an index, with 1995 = 100, 1973 = 106 (value added)[5] or 92 ('output')[6] was the highest peak reached by either series until 1989–90. Output in 1973 is again shown as still dominated by new work (£30 billion/£48 billion or 62% at 1995 prices), unlike today (50%).

These are years of prices and incomes policies, designed to permit the economy to operate at something close to full employment without provoking accelerating inflation. The gross construction price index (1995 = 100, used by DETR to deflate construction output to constant price volume) stood in 1973 at 19, having risen from 12 in 1969 (cf. 7 in 1955). In 1973 the index of construction materials prices rose by 6% and of labour costs by 13%; in 1974, materials prices rose by 32% and labour costs by 15%; in 1975, materials prices rose by 27% and labour costs by 21%. These are the highest rates of construction price inflation yet recorded for the UK. Moreover, they are far from being merely a reflection of some broader, non-construction-related inflationary phenomenon. The sectoral inflation differential, construction price inflation minus the gross domestic product (GDP) deflator, hit all-time peaks in 1973 and 1974 of around +15% per annum. Tender prices for local authority (LA) housing rose by 45% between the third quarter of 1972 and the third quarter of 1973. This unprecedented inflation was the context in which the research study was done and the report written. It had not, however, been such an obvious issue when the terms of reference were initially drawn up.

These are years before the big surge in the number of very small construction firms. In 1973 there were 96 000 firms on the DOE register (cf. 165 000 in 1999 and 210 000 in 1990).

1972 saw the last British national building strike of the twentieth century, partly around issues of pay but also around issues of union recognition on sites and in protest against firms' use of 'the lump'. Self-employment had begun to grow and direct employment to fall. The Phelps-Brown inquiry into this 'new' phenomenon had been published in 1968[7]. Although the series for total output (including the estimated output of unrecorded small firms and self-employed) was still rising (from index 90.9 in 1964 to 107.6 in 1973), direct employment of operatives by contractors had peaked as early as 1964 at index 129.8, and by 1973 had fallen to 99.3. Part of this fall might be attributable to real labour productivity in construction rising faster than output volume, but most will be attributable to the switch towards self-employment and 'the lump'. Selective Employment Tax had been introduced in 1966, and the Redundancy Payments Act in 1965.These together significantly increased the cost to an employer of direct employment. Labour-only subcontracting also afforded a way around the wage rate restrictions of prices and incomes policies. Construction industry (contractors plus DLOs) direct employment was nearly 1 500 000[4].

Demand for labour in the construction industry had, by these years, been growing year on year for a long time. In 1973 there were only 85 000 registered unemployed construction workers, and only 15 000 of these had craft skills. The U/V ratio (the registered number of workers unemployed divided by the number of unfilled vacancies notified to employment exchanges) in construction had fallen from around 7.7 in 1969 to around 2.9 in 1973.

The construction sector had reached and just passed a secular peak in its share in UK GDP, with completion of elimination of the post-war building backlog, and the first signs of the onset of deindustrialisation[8] and of decline in the condition of the public infrastructure. The 1973 value added of the construction industry (narrow definition, i.e. excluding DLOs and construction professional service firms) plus that of the building materials industry equalled 10% of GDP according to the Wood report. According to the ONS[9], construction output as a percentage of GDP was at its all-time peak of 11.5%, compared with 7.5% in 1995 and 10% in 1989.

From 1957 to 1971, GDP grew steadily at 2% per annum (never faster than 3.5% or slower than +0.5%). Since 1971, GDP growth has become much more unstable (peaks of up to 7%, troughs of up to –3%), while continuing at much the same trend rate, 2%. Deflated construction output had, at the time of the Wood report, been flat since 1967 (1973 was only just higher than 1967, and the years between lie below these twin peaks). From 1957 to 1967, in contrast, construction industry (CI) output (DOE definition) had been growing at over 5% per annum on average (and thus faster than GDP), compared with negligible (below 1%, and thus slower than GDP) average per annum growth since. Moreover, the representatives of the construction industry were about to experience the fate of the boy who cried wolf. Up to 1973 this growth had in fact been fairly stable (a 1 year maximum of 9%, and a minimum of 2%). Nevertheless, throughout this period, up to and including the Wood report, the industry had complained repeatedly that its weaknesses and inefficiencies were all attributable to its having to face an exceptionally unstable demand level[10]. Since 1973, CI output has in fact become much more unstable (maximum of +11.5%, minimum of –11%.), but, by the time the wolf finally did appear, no one was paying attention.

The year 1974 was one (the first) of three great recessions in CI output since the war (–11% in 1974; –9% in 1981; –9% in 1991), corresponding to three major GDP recessions and troughs (–2% in 1974; –1.5% in 1981; –3% in 1990).

8.2   Summary of the report

The report has two main themes: reforms of public sector procedures to yield better value for money from each public sector project and from capital budgets; and proposals for more active management of the volumes of public sector capital expenditure to yield greater stability and predictability of demand, and thus permit the industry to take the kind of fixed-cost-increasing measures that would improve its efficiency and performance in the longer term.

The second set of proposals was launched, already 'dead in the water', into a context of worsening economic crises and unplanned and drastic cuts in public expenditure. For the first set of proposals the time was in one further respect unpropitious. The Poulson scandal had just been investigated, revealing an association of corruption of local government officers and councillors by the owners of construction firms with the practices of 'package deals' and 'design and construct' that the report sought to promote:

> 'Current anxieties about conduct in local government might also lead to the introduction of rigidity into tendering and contract procedures. We argue …that neither public accountability nor considerations of value for money demands an inflexible approach. In our view procedures which tend to produce continuity of work for designers …and the construction industry demand more general adoption.'

The Wood report aimed to solve the difficulties characteristic of the 1960s, arising from discontinuity of public construction expenditure (minor instabilities of workloads around a sharply rising trend) by recommending each spending authority to develop (and be allowed by the Treasury to keep to) a rolling programme of at least 3 years of construction spending. Its solutions are: fixed targets for (steadily growing) public capital expenditure, to be adhered to in spite of macroeconomic conditions; more serial and continuity contracts, pooling of work by smaller clients into consortia and treatment of projects as part of planned programmes, not as one-off; and more care and resource to be given to the client briefing, designer selection and project management processes:

> 'Our chief anxiety is that value for money may be defined too narrowly in that the desirable aim of achieving, contract by contract, the lowest priced work is allowed to obscure the need to ensure that the industries are able to use their resources most effectively and, above all, encouraged to improve their efficiency and performance in the longer term. The key to achieving these aims lies in providing greater continuity of work …This means, at central govt level, avoiding …actions which might accentuate variations in demand (i.e. cause public and private sector demand peaks

or troughs to coincide) and, at the level of the individual client, assuring both designers and contractors of a more secure and predictable (public sector) workload.'

Why will continuity of public sector workloads for particular firms yield such beneficial results? Wood's answer is in two parts, the first applying to designers and the second to contractors.

For designers, Wood found that there were many similar small projects, mostly designed by clients' in-house design teams. A resource/demand balancing problem then arose, in which consultants were used partly for larger, complex projects but partly for 'overflow' demand. Wood finds three modes of design provision: in-house departments; contractor design as part of design-and-build packages; and consultant design. Selection of the mode of provision of design for a project or set of projects Wood found to be handled without obvious rationale in many cases. Consultant selection was found to be very ad hoc (never, of course, on fee competition, but not usually on any form of competition). Poor (absent) design management or coordination was found in many cases, even on larger projects.

A sense comes over of design departments and consultants both overstretched by the number and value of project demand relative to their design resources. Hence, Wood's stress on predictable workloads, to enable them to do some manpower planning and training.

For contractors, Wood found that the traditional system was no longer working as it was supposed to. Building projects were almost always tendered for lump sum price with supposedly full Bill of Quantities, but in fact usually with incomplete design. The value of alterations made after contract agreement averaged 10% of the contract value. In only 39% of building projects were 'all' drawings and information available at tender stage. Incomplete design is attributed particularly to lack of resources within in-house departments at times of boom in demand, and to lack of effective project management. Two possible recommendations are discussed as means to deal with the incompatibility of lump sum tendering, including selection on price, and incomplete design at tender stage: use another method of contractor selection and ensure designs are complete at tender. The Wood report urges the latter as preferable, except for very large and complex projects, where it recommends alternative modes of selection, to get contractor input into design, or for 'projects, large or small, where there are opportunities for allowing more continuous working by contractors to be seized'.

Wood advocates contractor design (design and construct) for more of the simple, potentially repetitive projects.

Wood found that, 10 years after Banwell[11] and 30 years after Simon[12], open competition was still used to select the main contractor in 16% of projects. Select competition was found to predominate. Wood urged greater use of alternative methods (two-stage tender, serial/continuity tender and negotiation) and 'formulation of more balanced, but not less rigorous, concepts of accountability'. On the working of the system of select lists of approved contractors, Wood emphasised the need for regular updating and recommended

> 'restricted lists predominantly of firms with a record of satisfactory work for the client. In this way the client benefits from and provides an incentive to the maintenance of good working relationships.'

Wood also recommended enquiries as to current resources and workload before a list of firms to be invited to tender is drawn up.

Continuity, as opposed to one-off projects, would give 'the advantages gained through learning particular operations and sequences, understanding client attitudes and policies, and welding together effective management and site teams'. Its absence 'inhibits training effort and .. the incentive to innovate is weakened'.

Problems of reconciling recommended best practice in procurement with the EU Directives on public works and competition were already (1975) seen as a potential difficulty. Similarly, the recommendation that the problem of smaller, inexperienced clients lay in the pooling of their projects (cf. Partnerships UK) has a distinctly contemporary resonance[13].

8.3   Review of the report's impact

In effect, the Wood report is a true 'successor' to Banwell, working with a similar diagnosis and differing mainly in placing a relatively greater emphasis than Banwell had done on the importance, for value for money, of the briefing and design stages as opposed to contractor selection. It fills in some omissions of Banwell and, had it appeared in 1965 rather than 1975, might have had a high, and beneficial, impact. Wood identified, for instance, the way that the 3 year rolling programme capital expenditure systems of the Department of Education and Science (DES) had permitted development of consortia such as CLASP and SCOLA. These are commended and seen as the way forward, whereas in fact they were to fall early victims of the approaching wave of public sector cuts and retrenchment.

This is a classic example of a report left behind by changes in the macroeconomic and political-economic context. Wood's recommendations had some modest impact in the period 1975–9[14] but were made irrelevant in 1979, which year saw a new Government only interested in cutting public sector construction expenditure, not in using it as a lever to shape the CI. Value-for-money (VFM) tests became for a while a kind of 'code' or synonym for expenditure controls and reductions, rather than meaning, as Wood had seen them, how to get more output from the same level of expenditure ('obtaining the maximum within budget').

Though the Wood report emerged into the stagflation crisis of the mid-1970s, the report took no direct notice of what had happened since 1972. It presumed a future return to the 'normalcy' of the 1960s, not foreseeing that this golden age was gone, never to return. Its aim was to solve the minor difficulties of the 1960s: inappropriate macroeconomic 'fine tuning'; difficulties of planning for growth by firms, because of minor instabilities of workloads around a sharply rising trend; and difficulties of planning for growth in capital spending by Government clients, because of lack of rolling programmes, lack of 3 year expenditure budgets and lack of project management expertise. It had no suggestions to offer to the new world of retrenchment and cuts.

At the level of project procedures, the Wood report appears as rather more prescient – even, in some respects, 20 years ahead of its time.

In terms of proposing solutions for the (potentially recurrent) problems faced by public sector projects in particular at times when the industry at large is stretched to, or beyond, its capacity, Wood did not face up to the idea that incomplete design at tender may be unavoidable, in the context of a system of annual budgeting (where contracts must be awarded in the fiscal year to which budget has been allocated) and of shortages of design and project management resources and of having to compete for staff with private clients' projects. The problem of consequent claims, variations and overruns is identified, as is the potential for what we would now call 'adverse selection' of contractors. Wood suggests (p. 46) that standard select competitive tendering but on approximate bills of quantities may be sufficient remedy to the

> 'multifarious consequences of this abuse of the orthodox approach .. evident in the high level of variations, claims and even disputes on such contracts.'

However, the argument given for advantages for approximate bills is weak, and the sources of the problem are not rigorously analysed.

8.4   Conclusion

It is surely unfair to blame the members of the Wood Committee for lack of foresight (which contemporaneous studies did any better in this respect?) but equally impossible to read it now except through the lens of hindsight. These particular times, truly, were out of joint.

There may be lessons to learn from the composition of this Committee. Central government representation was limited to the head of DOE's Construction Directorate and the Department of Health and Social Security (DHSS). None of the bodies to which recommendations for action were addressed (the Treasury; local government directorates of the DOE; the Property Services Agency; other spending ministries; consortia of local authorities) were represented.

It is interesting to note that Sir Kenneth Wood, who was Chairman of Concrete Ltd, only replaced Mr David Morrell, Chairman of Mitchell Construction, as Chair of the Working Party because the latter resigned in 1973. Readers are urged to look at Morrell's book, *Indictment*, for the fascinating background to that resignation.

The report arose, perhaps uniquely among those reviewed in this volume, in response to an initiative by trade union members of the economic development committees (EDCs). They asked whether the power potentially offered by the massive size of the public sector purchasing programme as a whole could be used more actively: to reorganise the industry (fewer firms, each with continuous and planned workloads from the public sector, each able to invest more in production-cost-reducing technologies and each presumably offering continuity of employment) and at the same time to reorganise the project process (to reduce problems of time overruns and high levels of variations, claims and disputes).

There is some ambiguity throughout the report as to whether to focus on reorganisation of the industry or on dealing with malfunctions in the project process. The reconnaissance survey of CI firms and public sector financial and professional officials generated a list of complaints mainly by firms about adverse effects of various public sector practices on them. This list then shaped the research study. Although the main report does develop an argument in which the implied 'complainant' is Government and its customers ('why can't the CI be more efficient and give us better VFM'), it does so only timorously, and without the backing of research investigation into the practical alternatives it concludes by recommending (for example, research studies of pilot client consortia, or of serial/continuity agreements in use). There is therefore some disjunction between the focus of the research study (Appendices A and B) and the focus of the Summary and Recommendations (Chapter 7). Moreover, the lack of evidential support for its proposals (in the form of 'demonstration' studies) and its vagueness about where and when continuity contracts would be appropriate cannot have helped it have an impact on public sector clients and their advisors.

In one sense, the Wood report came up with a case for partnering and prime contracting, *avant la lettre*, over 20 years before these would become the fashion.

The recommendations for Government to pull together, announce and fix over a 3 year plan horizon all public sector construction expenditure intentions were eventually to be met at least by advanced publication of such planned figures, even though plan revisions and discrepancies between plan and outturn remain the norm.

The recommendation that a project requires a sole and senior client representative, with sufficient expertise, authority and time, and that public officials may need advice, support and training to enable them to play this role effectively could be found repeated, today, in Office of Government Commerce guidance – a clear sign that a problem recognised is not necessarily a problem solved.

8.5   References

1   Marglin, S. & Schor, J. (1990) *The Golden Age of Capitalism: Reinterpreting the Postwar Experience*. Clarendon, Oxford.
2   Armstrong, P., *et al.* (1984) *Capitalism since World War II*. Fontana, London.
3   Direct Labour Collective (1978) *Building with Direct Labour*. Conference of Socialist Economists, London.
4   DOE (1974) *Housing and Construction Statistics, No.9*. HMSO, London.
5   ONS (2000) *United Kingdom National Accounts*. The Stationery Office, London.
6   DETR (2000) *Construction Statistics Annual*. Department of the Environment, Transport and the Regions, London.
7   Phelps-Brown, E. (1968) *Report of the Committee of Inquiry into Certain Matters Concerning Labour in Building and Civil Engineering*. Command 3714. HMSO, London.
8   Bon, R. & Crosthwaite, D. (2000) *The Future of International Construction*. Telford, London.
9   ONS (1997) *Economic Trends: Annual Supplement*. HMSO, London.
10 Sugden, J. (1975) The place of construction in the economy. In: *Aspects of the Economics of Construction* (ed. D. Turin). Godwin, London.
11 Banwell, Sir Harold (1964) *The Placing and Management of Contracts for Building and Civil Engineering Industries*. HMSO, London.
12 Simon, Sir Ernest (1944) *The Placing and Management of Building Contracts*. HMSO, London.
13 Construction Industry Council (2000) *The Role of Cost Saving and Innovation in PFI Projects*. Telford, London.
14 Ive, G. (1983) *Capacity and Response to Demand in the Housebuilding Industry*. UCL, London.

Chapter 9
Faster Building for Industry:
NEDO (1983)
John Connaughton & Lawrence Mbugua

9.1 Overview of the economic, political and social scene 1978–83

9.1.1 The UK economy

Faster Building for Industry[1] was published in 1983. That year, the Conservative Party won the general election with the most decisive majority in 30 years; Norman Foster won his first major London commission; and a *World in Action* programme severely dented public confidence in timber-frame housing.

The late 1970s and early 1980s were times of considerable economic upheaval in the UK. For more than 20 years before the mid-1970s, output (excluding oil) had risen fairly steadily (2.5–3% per annum). However, around 1973/4 output growth started to fall, and in 1985 it was about 20% below trend[2]. This was a consequence of falling demand after 1973, which also caused unemployment to rise sharply. By the mid-1970s, UK unemployment had risen above 1 million for the first time and continued to grow to over 3 million by the mid-1980s.

These trends were not felt equally throughout the country. The 'north–south divide', the disparity in investment and wealth between the northern and the (wealthier) southern parts of the UK, widened during this period and continued to do so during the 1980s.

9.1.2 The construction industry

During the late 1970s and early 1980s, the construction industry accounted for a declining proportion of gross domestic product (GDP). The peak was in 1972 at 8% of GDP (measured at factor cost), declining to under 6% in the mid-1980s.

In the 5 years before *Faster Building for Industry*, the construction industry was the single largest employer among the production industries, with about 10% of the nation's labour force employed either directly in construction or in the associated professions and suppliers. The gross output of the industry represented 8% of GDP and also accounted for half the nation's fixed capital investment. By 1980 there were 224 000 unemployed construction workers, representing a 15% unemployment rate, though this had increased by almost 50% from the previous year.

Over the period, there were significant changes in the sectoral distribution of construction output. Work in the commercial sector grew by some 20% in real terms in the period 1978–83, whereas public expenditure on new housing declined by some 50%. In the industrial sector, output fell by some 30% in the period, and by 1983 work on industrial building accounted for some 7% of all construction output[3] (see Table 9.1).

Table 9.1 Construction output at 1985 prices (in £ million).

	1978	1983	1988	1993	1998[r,p]	2000[p]
Repair and maintenance (R&M)						
All R&M	10 839	12 220	14 670	12 174	13 499	13 437
New housing						
Private	4 586	4 174	5 312	3 136	3 648	3 670
Public	3 077	1 209	789	1 185	736	823
Other new work						
Commercial	2 640	3 172	5 395	4 321	6 630	7 759
Industrial	2 939	1 982	3 463	3 478	5 056	4 704
Public	4 713	3 953	3 369	5 114	4 204	4 587
Infrastructure				7 480	6 420	6 464
Total	**28 794**	**26 710**	**33 998**	**36 888**	**40 193**	**41 444**

[r] revised; [p] provisional. Source: DETR, *Housing and Construction Statistics*, (various).

9.1.3 Recession and response

1980 was a particularly difficult year for the construction industry as markets both in the UK and overseas withered in the face of high interest rates, high exchange rates and an industrial recession caused by the 'oil crisis'. The global recession was further deepened by the domestic downturn in the housing market and severely constraining Government policies. Though larger firms, with an annual turnover in excess of £25 million, weathered the storm rather better than their smaller competitors – helped by their involvement in overseas work – both had to contend with cutbacks in domestic public expenditure and a declining construction market generally.

The bottom of the recession was reached around the turn of 1980–81, when many property developers had failed to sell their investments, and bank repossessions were not uncommon. Inflation and interest rates had declined steadily to 1982. This contributed to a rise in the value of capital stock – houses, shares, etc. – and thus provided a stimulus for consumption in late 1981 and 1982.

In response to successive reductions in public sector construction expenditure and other cutbacks associated with the International Monetary Fund (IMF) loans in late 1976, a group of representatives from trade unions and professional institutions got together in 1977 to form the Group of Eight (G8). Its aim was to provide a voice for the industry and, in particular, to ensure that Government addressed its concern that construction was bearing a disproportionate share of cuts in public expenditure. It was especially concerned about the long-term consequences of this decline on the industry's future efficiency. In 1981, G8 consisted of the Institution of Civil Engineers, the Royal Institution of Chartered Surveyors, the Construction and Civil Engineering Group and Building Crafts Section of the Transport and General Workers Union, the National Council of Building Material Producers, the Royal Institute of British Architects, the Federation of Civil Engineering Contractors, the National Federation of Building Trade Employers and the Union of Construction Allied Trades and Technicians.

G8 was, for a while, the main channel through which the Government engaged in dialogue with construction on matters of strategic importance to the industry, although it was not a separate organisation per se (i.e. it had no staff or formal constitution of its own). Its critics considered it to be unrepresentative of the industry as a whole and an ineffective lobbying body in conveying the industry's plight to Government. It was disbanded in 1987, and construction has not, since then, had a sole body to speak to Government on its behalf.

9.1.4   *The industrial scene*

The late 1970s and early 1980s also witnessed significant changes in the industrial sectors of the economy. A prolonged period of deindustrialisation, starting in the early 1960s, was coming to a conclusion. While employment in manufacturing fell from some 8 million in the 1960s to some 5 million by the mid-1980s – as Britain's traditional manufacturing sectors steadily declined – more than 1 million manufacturing jobs were lost in the recession of the early 1980s[4]. Government attempts to effect industrial restructuring were dominated by policies aimed at encouraging industry away from heavy manufacturing and port-related activities and towards the newer, information-based 'high-tech' industries.

In the mid-to-late 1970s, the National Economic Development Office (NEDO) perceived that Britain's industrial productivity was being hampered

by outmoded premises; in addition, the construction and engineering industries were also in decline. A programme of industrial building would not
only assist in the regeneration of industry, it was felt, but would also help
improve the fortunes of the construction industry[5]. During this period the
provision of new industrial premises by the public sector was a well-estab-
lished aspect of policy, and such building programmes tended to be stepped
up when economic growth faltered. However, the approach began to fall out
of favour in the early 1980s as the Conservative Government developed policies that challenged the notion of direct public participation in what it saw as
a purely private sector responsibility (see Fothergill *et al.*[4], Chapter 6, for a
discussion).

It is against this background of industrial decline and public policy shift
that *Faster Building for Industry* was published in 1983.

9.2   *Faster Building for Industry* – background and summary

9.2.1   *Background and purpose*

In contrast to *Construction for Industrial Recovery*[5], *Faster Building for Industry*[1]
was focused more clearly on how the private sector should help itself. It
contained little by way of policy recommendations. In essence it reflected the
concerns of business – that the process of industrial building in the UK was
slow and burdensome – and offered advice on how to improve the procurement process. These concerns were high on the agenda during the late 1970s. A
study by Slough Estates[6], investigating the problems of developing a 'typical'
50 000 ft^2 factory, found that development was, in almost every way, more
difficult in Britain than in other European countries and North America (see
Table 9.2). Additionally, studies by the Centre for Advanced Land Use Studies
(CALUS)[7] and the Building Research Team[8] argued for improvements in the
kind of premises offered to industry by the development-for-rent sector.

Table 9.2 International comparisons of construction performance 1977–8.

	UK	Canada	Belgium	USA	France	Germany	Australia
Preparing drawings (weeks)	20	4	6	6	9.5	12	14
Planning approval (weeks)	26	6	6	4.5	16	12	3
Building time (weeks)	57	21	37	23	30	29	28
Costs index (UK = 100)	100	59	107	74	98	87	94

Source: Slough Estates[6].

The (then) Department of the Environment, headed by Michael Heseltine, funded the Building Economic Development Committee (EDC) of NEDO to investigate claims that the process of obtaining industrial buildings was more troublesome in the UK than elsewhere, and to make recommendations on how it could be improved. The Building Research Establishment (BRE) led the research and, supported by Kenchington Little and Partners, sought to address the key EDC objectives:

> 'to establish the key factors that affect the time taken to construct industrial buildings, to identify best practice in planning and maintaining control of construction periods and to formulate recommendations for action by client, designer, and contractor.'

The intention was to seek both objective data and informed industry opinion to investigate construction and procurement processes. The emphasis on speed of construction/procurement was viewed as part of a wider concern about the effectiveness with which construction could meet client's needs, within budget and of the specified quality. The investigation concentrated not only on building project duration but also on the overall service clients receive from the construction industry, observing how, as a result of actions, relationships and communications between project parties, the construction process could be made more effective.

9.2.2   Methodology

The research involved a combination of quantitative and qualitative methods:

- A large-scale questionnaire survey of some 5000 projects built between 1980 and 1981 was undertaken. This provided quantitative data on project duration and other variables for statistical analysis, and also allowed some generalisation of findings from the more detailed, qualitative case studies of individual projects (see below). At that time, some 8000–9000 industrial building projects costing more than £10 000 were started each year.
- Detailed case studies of 56 recent industrial building projects (involving document analysis and interviews with key project parties) were undertaken to provide an improved understanding of the processes involved.
- Less detailed studies of another 114 projects due to start in 1980/81 were also undertaken.

9.2.3   Overall results, key findings and recommendations

The report's key conclusions were on matters that could aid speed or cause delay prior to construction as well as during construction, and its recommendations were addressed to both the industry and its clients. On the key issue of project duration, the study found little support for the idea that speedier projects carried penalties for clients in terms of extra costs, loss of flexibility or quality. The study concluded that construction was as fast as clients demanded from the industry. Rarely did technical or physical constraints pose limitations on the speed of construction. In the absence of any generally accepted times for given types of project, the onus for requesting 'fast' times was left to clients.

While a variety of procurement methods were available to clients, few had the know-how and experience to demand 'fast' times, to make informed choices about the organisation of their projects or to commission them in ways that guaranteed all their requirements would be met. Clients were not a homogeneous organised group and thus were not in a position to apply uniform standards, particularly as there were no easily accessible sources of objective and practical information that could guide their expectations and moves. Inexperienced clients were potentially disadvantaged; there was a significant opportunity for industry to fill the management gap arising from customer inexperience. The report suggested a role for a 'principal adviser'.

The study noted that clients were always concerned about the speed of their projects. It highlighted both 'good practice' and the factors that could interfere with the speedy and effective progress of projects (see below), noting that most were within the control of the project participants. In summary, fast projects required:

- knowledgeable and well-advised clients, who would provide substantial management input as well as demanding fast building without sacrificing either cost or quality;
- a clear and understandable design that took into account design contributions from specialist consultants and contractors at the appropriate time, as well as the practical aspects of organising work on site (i.e. 'buildability') and procuring materials in advance where appropriate;
- choice of the main contractor on the basis of quality and management capabilities as well as price;
- a coherent management control regime with clear responsibility for project progress, and project arrangements that allowed early, precise and integrated procurement, including adequate preparation for construction;
- high levels of site management with good back-up from the contractor's organisation, good communications with design team and client, and a clear and a detailed programme, backed by incentives (where appropriate), used actively to chase progress;
- the avoidance of contract variations where possible as they often resulted in delays;
- prior investigation of ground conditions.

The study also found that, while traditional methods of design and tendering can give good results, 'non-traditional' procurement (e.g. design and build, management contracting) tended to be more conducive to faster delivery. With traditional methods, both tendering on approximate bills of quantities and choice of contractor with a negotiated tender led to faster progress.

The study emphasised the value of cooperation and good communication between the participants, noting that an attitude of cooperation was conducive to fast construction times.

Finally, the study identified that the main external cause of delay was the inflexibility and inefficiency of statutory authorities. There were bureaucratic complications and a general lack of urgency in gaining planning and building regulation approvals, especially with regard to fire regulations, affected progress on over half the projects.

In summary, the study showed that the construction industry could deliver quickly and efficiently in the right circumstances. This was true both for the 'traditional' arrangements, defined as those in which the client had separate relationships with a design team and a contractor, and non-traditional arrangements, such as design and build contracts. However, the range of performance found in practice was very wide, both in the planning (pre-construction) period and in the time taken for construction itself. For example, the average construction time for buildings costing around £0.5 million was 9 months, but 10% took less than 6 months and 10% more than 13 months.

The research identified substantial scope for improving the general pace of construction of industrial buildings without sacrificing quality or increasing cost. Moreover, it identified significant potential for improvement in the process of design and construction. While the prime responsibility for achieving these improvements lay with the industry itself, clients also had an important role in improving standards. In particular, the extent to which they could identify their requirements and articulate these to industry – as well as engage in the design and construction process to ensure these were met – was seen as central to improved construction performance.

9.3   Impact of *Faster Building for Industry*

9.3.1   *Introduction and overview*

The two key concerns giving rise to *Faster Building for Industry*:

- the importance of manufacturing industry to the UK economy
- the need for speed and efficiency in the process of construction

have very different relevance today than in 1983. In the following assessment of the report's impact, we concentrate on its legacy for changes in the *process of construction*. Nevertheless, it is worth noting that the significance of manufacturing to the UK economy has declined since 1983 – manufacturing's share of GDP has fallen steadily from about 30% in the late 1970s to about 20% today[9]. The extent to which policies for growth and improvement in international competitiveness have focused on regenerating the manufacturing base has also changed markedly over the period, with the dominant policy thrust now towards the service sector. So did the report give rise to improvements in the speed and efficiency of the construction process? Certainly, construction has seen the introduction of more accelerated construction processes, supported by new management and working arrangements, since 1983, and we will discuss how the report and subsequent developments may have influenced this. However, it is also in the report's *analysis* that we see its particular relevance to contemporary construction concerns, and perhaps its more durable legacy. Five key themes emerge from the report that are at the centre of current attempts to re-engineer the construction process:

- the focus on construction *customers,* and the extent to which they can *pull* the process in desired directions;
- the emphasis on *buildability*, expressed more in contemporary analysis as the integration of design and construction;
- the importance of *culture* – including personal and corporate attitudes and behaviour – in project delivery;

and, to a lesser extent

- the need for quality-based selection (of contractors and other project participants);
- the emphasis on quality (of the construction product), in addition to speed of delivery.

These key themes add up to much of today's analysis of shortcomings in the construction process and recommendations for improvement[10, 11]. That is not to claim undue influence or prescience for the report, for it should be viewed in the context of a larger body of work that emerged during the 1980s (including, for example, *Faster Building for Commerce*[12] reviewed elsewhere in this publication) concerned primarily with improving the performance of construction. Nevertheless, the report does preview these contemporary themes in a coherent manner, and perhaps its real legacy is in reminding us that these issues need to be brought together and 'joined up' if we are to make real progress with improving the overall process of construction. We will return to this point later, but first we will review briefly some of the changes in construction duration and organisation since *Faster Building for Industry*.

9.3.2   *Faster building …or more certain delivery times?*

As the industry moved out of recession in the early 1980s and into a period of high growth in the commercial and retail sectors in the mid-to-late 1980s, pressure mounted for faster construction times. Commercial property developers in particular demanded accelerated, 'fast-track' construction programmes to reduce time to market, and these necessitated new procurement and organisational arrangements to deliver them. It is clear from *Faster Building for Industry* that construction was already providing non-traditional arrangements such as management contracting and design and build. The popularity of USA-style construction management in the 1980s was influenced largely by developers keen to overlap design and construction and to share construction risk in order to achieve rapid and effective project delivery.

The Royal Institution of Chartered Surveyors (RICS) undertakes periodic surveys of building contracts in use and, while this is restricted largely to information provided by its members, it provides a useful indication of changing procurement and organisational trends in construction[13]. In Table 9.3, for example, we can see a fairly steady decline in traditional tendering between 1984 and 1998 (the latest year for which data are available) and a commensurate rise in design and build (and, to a lesser extent, management forms of procurement). While the need for faster construction is undoubtedly a key factor in these changes, other parameters – such as the need for cost certainty and contractual 'access' to key parties in the supply chain, for example – are also influential, and it is difficult to attribute changes in client preferences directly to the need for speed or any other requirement. Nevertheless, non-traditional procurement methods tend to be chosen because they are faster and tend to deliver faster times[14, 15].

Table 9.3 Trends in procurement methods – by value of contracts.

Procurement method	Percentage of contracts by value			
	1984	1989	1993	1998
Lump sum – firm BQ	58.7	52.3	41.6	28.4
Lump sum – specs and drawings	13.1	10.2	8.3	10.0
Lump sum – design and build	5.1	10.9	35.7	41.4
Remeasurement – approximate. BQ	6.6	3.6	4.1	1.7
Prime cost plus fixed fee	4.5	1.1	0.2	0.3
Management contract	12.0	15.0	6.2	10.4
Construction management	—	6.9	3.9	7.7
Total	**100.0**	**100.0**	**100.0**	**100.0**

Note. Percentages adjusted to exclude 'other contracts'. Source: RICS[13].

So has the process of construction become faster since *Faster Building for Industry*? We are not aware of any comprehensive survey of the duration of industrial building projects carried out since 1983 that would enable us to address this directly. Work on project duration, surprisingly enough, tends to concentrate on identifying the parameters affecting time performance[16] or on time forecasting techniques and modelling[17]. Large-scale surveys of time performance of construction projects appear thin on the ground. Our own firm has undertaken a survey of UK local government building projects (122 projects)[18, 19] with a mean construction period of 6 months. A similar survey of 186 local government projects undertaken by BRE some 15 years earlier in 1980[20] has a mean construction period of 14 months. In this, as in any comparison of project information on time performance, there are considerable definitional and measurement difficulties (whether to include preconstruction processes, how to define 'design' and 'construction' periods, how to deal with delays, how to normalise data for different project types, etc). While the faster times identified in the later survey are not at all conclusive, we believe they are indicative of more general improvement in time performance.

The pattern of construction procurement is quite different today to that in the early 1980s, and it is highly likely that, as a consequence, design and construction times have reduced over this period. However, speed (of design, construction or both) may no longer be a critical issue for UK construction. Certainly, construction continues to be criticised for its time performance, but the emphasis is no longer on speed (or lack of it), but on *lateness of delivery*.

A recent study of Government procurement[21] suggests that some 70% of Government construction projects are delivered late. Additionally, the Construction Best Practice Programme now provides data on a range of key performance indicators[22], including time performance – measured as time 'predictability' (another term for 'lateness', i.e. the difference between expected or planned performance and actual achievement). Latest figures (for 2000) show a slight worsening in overall project time predictability from 1998 to 2000. It will be interesting to observe whether future improvements in predictability are at the expense of speed of design/construction. Recognising the danger, and learning at least some of the lessons from *Faster Building for Industry*, perhaps the real challenge now is to set targets for faster, more certain design and construction times – and to achieve them.

9.3.3 Contemporary concerns

It is worth emphasising the extent to which the five key themes identified in the report have resonated throughout later attempts to set an agenda for industry improvement. This is not only restricted to UK construction[10, 23] but is evident in analysis of shortcomings in the construction industries of other countries also[24–26] (brief summaries of key industry reviews are provided by the National Audit Office[11]). In passing, it may be observed that such countries all have some elements of 'British' construction practice and organisation, but the extent to which these may be responsible for the shortcomings identified is a wide-open question, though an interesting one for research.

The more useful point is that these issues can perhaps be viewed as some of the 'universal' themes of construction management research. The Tavistock Institute, for example, has a long history of enquiry into social processes in construction, from Higgin & Jessop in 1965[27] through to Nicolini *et al.* in 2001[28], which emphasises the role of the client in the process and the importance of culture. (Other chapters cover many of these issues.) In the following paragraphs we want to focus on two related themes that seem to us to reflect the enduring legacy of *Faster Building for Industry*: the role of clients in the process, and supplier and process integration.

Clients – their nature and role

The focus on the 'customer' is now so pervasive in current business thinking that it has become commonplace in non-business discourse also. In *Faster Building for Industry* the customer focus is rather more tentative than in contemporary accounts, and the emphasis is on the distinction between experienced/knowledgeable customers and those that are less so – the latter are seen to be in need of more help and support from industry.

There is now a considerable body of guidance available to inexperienced customers to help them get the most from the construction process – for example, from the Construction Industry Board[29] and the Construction Round Table[30] – and guidance available from initiatives such as the Construction Best Practice Programme (www.cbpp.co.uk) is focused on the particular concerns of these customers. The larger question is whether this material is effective, though we are not aware of any relevant evaluation at the time of writing. We believe that the extent to which new customers can engage effectively with construction will continue to be an important challenge for the *industry* to address – for the uninitiated, the first critical point of contact with the industry is not at all clear, nor is the kind of advice that might be offered.

Perhaps the more significant development since *Faster Building for Industry* is that the more knowledgeable customers have been recognising the extent to which they can exert their influence over the construction process. At the core of ideas of lean production (and 'lean construction' in *Rethinking Construction*[10]) is the concept of 'customer pull'[31]. While this is viewed in ideas of modern manufacturing as a powerful, but not necessarily literal, force that provides impetus and focus to production, developments in construction suggest that customers are taking up the idea more literally.

Over the past 10 years or so, construction clients have been forming forums and organisations not only to share information and knowledge about the construction process but also to assemble their buying power to get a better deal from the industry. For example, the predecessors of the current

Confederation of Construction Clients, the Construction Round Table and the Construction Clients' Forum, both claimed to represent customers accounting for a significant proportion of UK construction work. Both confirmed their intention to use this potential buying power to drive industry improvement[32, 33]. The current Confederation now claims to represent the interests of construction clients collectively, with an aim of securing measurable and consistent improvement across industry (see www.clientsuccess.org). Government clients are now also represented collectively in the form of the Government Construction Clients Panel.

Customer intervention in the construction process is also evident in the extent to which many exemplar case studies of best construction practice are seen to be client-led. Cox & Townsend[34], for example, identify important cases of key corporate customers, such as Rover Group, McDonald's Restaurants and BAA, taking a lead in introducing many modern production principles, such as supply chain management and performance measurement, into construction projects. While *Faster Building for Industry* recognised the extent to which knowledgeable clients could get a better deal from construction, the gulf between them and their less experienced colleagues would appear wider than before. It is now commonplace for construction customers to be viewed as part of the construction supply chain, where their role is to participate collaboratively with suppliers, not only in the development of construction projects but in the creative resolution of problems[11, 35].

Process and supplier integration

Even the most proactive client in *Faster Building for Industry* would not have engaged in the supply chain to the extent that is now considered desirable. We believe that recommendations for increased client involvement in the process – in *Faster Building for Industry* and elsewhere – have helped over time to encourage clients to seek more collaborative ways of working.

The development and adoption of partnering is a relatively recent phenomenon in UK construction (*Partnering: Contracting without Conflict*[36] being one of the earlier references) and can be seen as something of a culmination of this trend. While there are many who exhort its benefits[37, 38], few have taken the time to examine its real impact, a recent review by Bresnen & Marshall[39] being one exception. It is certainly surprising how much it appears to be advocated in construction (though information on the extent of its adoption is not readily available) principally on the 'promise' of greater benefits rather than rigorous demonstration of how these can be achieved[40]. Bresnen & Marshall[39], for example, argue that there is little empirical research that would provide compelling evidence either for or against the adoption of partnering in different circumstances, though note that partnering – and, indeed, other more traditional forms of working – can provide benefits in certain situations.

The point here, however, is that collaborative working (of which partnering is perhaps a more particular form) is still in its infancy in construction,
and we have still much to learn about its benefits and methods. However, it is
helping us to understand what a more integrated design and construction
process – a key aim, if not an entirely clear vision, of *Faster Building for
Industry* – might look like. As customers and their suppliers work together
more closely, it is perhaps natural that they would seek to exploit such collaboration by reaching more deeply into construction supply processes – the
construction supply chain.

As already noted, the development and adoption of supply chain integration and management in construction has also been led, at least in part, by
clients who believe that their experience with rationalising lines of supply in
other sectors (such as manufacturing and retail) can be made to work in
construction too. While this development is also at a very early stage in
construction, initial experience is promising. The *Building Down Barriers*
project[28] identifies some of the potential benefits in terms of removal of waste,
errors and duplication and, more importantly, the integration of key
processes of design, supply and construction. Once again, much work
remains to be done to take these developments forward towards more effective, integrated design and construction processes. Though we now seem
quite a long way from *Faster Building for Industry*, we believe we can still feel
the pulse of some of its ideas. We now turn to draw some conclusions about
how these ideas may be taken forward.

9.4   Faster ...better ...or both?

While we have identified a good many contemporary construction themes in
Faster Building for Industry, the report's authors would have been remarkably
prescient had they consciously foresaw the changes taking place in UK
construction over the past few years. In our view this does not make the
report any the less interesting. *Faster Building for Industry* is an extremely
useful report for what it tells us about the past, and also for what it tells us
about the present. Considerable progress has been made on many of its
recommendations, though almost 20 years later we are still only starting to
understand some of its more challenging prescriptions for improvement.

We should take heart from this, rather then bemoan a lack of real progress.
Many of these prescriptions – whether foreseen or not – go to the core of what
we now believe can help improve the performance of our industry. Additionally, we are starting to see more clearly now (than we could have in 1983)
that engaging clients effectively in the construction process and integrating
design and construction processes have profound implications for how our
industry does business. The next step is surely to investigate these implications, for they will not be resolved simply by exhortation to do better.

It seems to us that an emerging awareness of the connections between many of the key themes we identified – customer engagement, the need to integrate disparate though vital design and construction processes, the importance of a supportive 'culture' of business – may lead to improved understanding of the forces shaping construction's performance, now and in the future. Blockley & Godfrey[41] in a stimulating analysis of the new thinking and management systems needed to move forward *Rethinking Construction*[10], speak of these connections and argue that 'everything is connected'.

What we think they mean, in part, is that some of the work we have tracked since *Faster Building for Industry* may shortly face an impasse, for it cannot realistically proceed without profound change in construction: change in the roles and functions of all the parties involved in construction projects; change in the social, organisational and commercial frameworks within which construction takes place; change in the very nature of what it means to build. At last we may be seeing more clearly what needs to be done to address the more fundamental challenges of *Faster Building for Industry* and much of what followed.

The next chapter looks at how subsequent work by NEDO built on the key findings of *Faster Building for Industry* and considered their implications for commercial development.

9.5   References

1   NEDO (1983) *Faster Building for Industry*. HMSO, London.
2   Coutts, K. J., *et al.* (1986) *The British Economy: Recent History and Medium Term Prospects*. Faculty of Economics, Cambridge.
3   DETR (2000) *Construction Statistics Annual, 2001 edn*. Department of the Environment, Transport and the Regions, London.
4   Fothergill, S., Monk, S. & Perry, M. (1987) *Property and Industrial Development*. Hutchinson, London.
5   NEDO (1978) *Construction for Industrial Recovery*. HMSO, London.
6   Slough Estates (1979) *Industrial Investment: A Case Study in Factory Building*. Slough Estates, Slough.
7   CALUS (1979) *Buildings for Industry*. College of Estate Management, University of Reading.
8   Building Research Team (1982) *The Small Advance Factories in Rural Areas: Final Report*. Department of Architecture, Oxford Polytechnic.
9   CSO (various) *National Accounts*. Central Statistics Office.
10   DETR (1998) *Rethinking Construction* (the Egan report). Department of the Environment, Transport and the Regions, London.
11   NAO (2001) *Modernising Construction*. National Audit Office, London.
12   NEDO (1988) *Faster Building for Commerce*. HMSO, London
13   RICS (2000) *Contracts in Use: A Survey of Building Contracts in Use during 1998*. Royal Institution of Chartered Surveyors, London.

14   Naoum, S. G. (1991) *Procurement and project performance – a comparison of management contracting and traditional contracting*. Occasional Paper no. 45, The Chartered Institute of Building, Ascot

15   Walker, D. H. T. (1995) An investigation into construction time performance. *Construction Management and Economics*, 13(3), 263–74.

16   Bromlilow, F. J. (1988) The time and cost performance of building contracts. *The Building Economist*, 27 September, 4.

17   Laptali, E., Bouchlaghem, N. M. & Wild, S. (1996) An integrated computer model of time and cost optimisation. In: *Proceedings of 12th Annual ARCOM Conference*, Sheffield Hallam University, 11–13 September, 133–9.

18   Audit Commission (1996) *Just Capital – Local Authority Management of Capital Projects*. The Audit Commission, London.

19   Audit Commission (1996) *Capital Expenditure – Audit Guide*. The Audit Commission, London.

20   Simms, A. G. (1984) *The final account and factors influencing it*. BRE Note 150/84, Garston.

21   Agile Construction Initiative (1998) *Constructing the Best Government Client. Benchmarking: Stage Two Study*. A report produced by the Agile Construction Initiative for HM Treasury and the Government Construction Client Panel, London.

22   DTI (2001) *Construction Industry KPIs Pack*. Construction Best Practice Programme/Department of Trade and Industry, London.

23   Latham, Sir Michael (1994) *Constructing the Team*. HMSO, London.

24   Construction Industry Council and Department of the Environment (Ireland) (1997) *Building our Future Together: Strategic Review of the Construction Industry*. Department of the Environment, Dublin.

25   Department of Public Works (South Africa) (1997) *Creating an Enabling Environment for Reconstruction, Growth and Development in the Construction Industry*. Department of Public Works, South Africa.

26   Construction Industry Review Committee (2001) *Construct for Excellence*. Construction Industry Review Committee, Hong Kong.

27   Higgin, G. & Jessop, N. (1965) *Communications in the Building Industry*. Tavistock Publications, London.

28   Nicolini, D., Holti, R. & Smalley, M. (2001) Integrating project activities: the theory and practice of managing the supply chain through clusters. *Construction Management and Economics*, January–February, 19(1), 37–47

29   Construction Industry Board (1997) *Constructing Success*. A code of practice for clients of the construction industry. Construction Industry Board/Thomas Telford, London.

30   Construction Round Table (1995) *Thinking about Building*. Construction Round Table, London.

31   Womack, J. P. & Jones, D. T. (1996) *Lean Thinking*. Simon and Schuster, New York.

32   Construction Clients' Forum (1998) *Constructing Improvement: The Clients' Proposals for a Pact with the Industry*. Construction Clients' Forum, London.

33   Construction Round Table (1998) *The Agenda for Change*. Construction Round Table, Building Research Establishment (BRE), Garston

34   Cox, A. & Townsend, M. (1998) *Strategic Procurement in Construction*. Thomas Telford, London.

35   Holti, R., Nicolini, D. & Smalley, M. (2000) *The Handbook of Supply Chain Management: The Essentials*. CIRIA, London.

36   NEDO (1991) *Partnering: Contracting without Conflict*. HMSO, London.

37   Bennett, J. & Jayes, S. (1995) *Trusting the Team: The Best Practice Guide to Partnering in Construction*. Centre for Strategic Studies in Construction, University of Reading.

38   Bennett, J. & Jayes, S. (1998) *The Seven Pillars of Partnering*. Reading Construction Forum, Reading.

39   Bresnen, M. & Marshall, N. (2000) Building partnerships: case studies of recent client-contractor collaboration in the UK construction industry. *Construction Management and Economics*, October–November, 18(7), 819–32.

40   Fisher, N. & Green, S. (2001) Partnering and the UK construction industry. The first ten years – a review of the literature. In: *Modernising Construction*. National Audit Office, London.

41   Blockley, D. & Godfrey, P. (2000) *Doing it Differently: Systems for Rethinking Construction*. Thomas Telford, London.

Chapter 10

Faster Building for Commerce: NEDO (1988)

Steven Male

10.1 Introduction

This chapter reviews the report *Faster Building for Commerce*[1], published in November 1988. It has to be read in the context of two earlier National Economic Development Office (NEDO) reports:

- *How Flexible is Construction*[2], published in May 1978, concerned with studying the resources and participants in the construction process and the ability of both the building and civil engineering industries to respond to changes in demand.
- *Faster Building for Industry*[3], published in June 1983, concerned with establishing key factors affecting the time taken to procure industrial buildings.

The 1983 *Faster Building for Industry* report[3] and the 1988 *Faster Building for Commerce* report[1] were undertaken as a package of research to identify why the process of procuring new industrial or commercial buildings was long, difficult and appeared to take a more time in the UK than overseas.

This chapter briefly places the *Faster Building for Commerce* report in the context of the economy, the political climate and the industry of the time. The report is outlined in terms of its structure; key themes are explored and a commentary is provided. Finally, the chapter concludes with a view of the *Faster Building for Industry* and *Faster Building for Commerce* reports in the context of Latham and Egan.

10.2 The report's political, economic and industrial context

10.2.1 *The politics of the Thatcher years and the economic environment of the 1980s*

The research for the report was conducted during the second term of Margaret Thatcher's premiership. The report was published during her third term of office. The 'Thatcher years', as they have become known, were characterised by a style and approach to politics that were distinctive, arguably producing a fundamental shift in the climate of the country. In essence, Thatcherism was an

> 'instinct and a series of values rather than a fully worked out ideology or set of policies. It is not synonymous with monetarism or a pure free market approach.'

Thatcherite values are derived from a belief in hard work, family responsibility, striving and the postponement of satisfaction, a sense of duty and patriotism, a dislike of trade unions, a promise to reverse Britain's post-war decline and a belief in a shift in the balance between the state and individual freedom. At one level, these reflected the mood of the times, on another, Mrs Thatcher's rise to power and continuation in power occurred because the electoral system produced significant majorities with a relatively small percentage of the vote, in fact less that half of the population voted her Governments into power.

The election of Mrs Thatcher in 1979 was more to do with the then Labour Government losing than with a positive shift to Thatcherism[4, p. 5]. It did, however, result in a political phenomenon that has yet to be repeated. The Thatcher years were about authoritarian but, when required, pragmatic Government; deep recession and recovery and then recession; privatisation and the transfer of public sector to private sector monopolies; wider share ownership; the fight against inflation and significant and stubborn high levels of unemployment – 3 million or 11% of the workforce in 1988; the miners' strike and the subsequent weakening of union power; the Falklands War; ever closer ties to the EEC masked by the very public fights over Britain's payments to the EEC and the special relationship with President Reagan; wider house ownership, including the purchase of council houses; the rise and fall of the SDP and the rise of the Liberal Democrat Alliance. The 1988 report was also published 4 years before the introduction of the 'free market' in Europe.

Brittan[5] concludes that an analysis of Thatcher's management of the economy during the period 1979–1988 poses a paradox. He argues that the special features of the 'Thatcher experiment' had their roots in the policies already in force under the previous Labour Government, or were being mirrored by events in other countries anyway. When looking at the evidence, he proposes that the Thatcherite, market-based policies of privatisation and suspending exchange controls were also to be seen in the context of interventionist policies in the car industry, in imports and in tax subsidies for home owners, complemented by the continued growth in the welfare state, around which any competent Premier would have to take action, regardless of political persuasion. He also suggests that the apparent overt weakening of unions by legislation was an attempt to give power back to the membership and away from the union 'barons'. The decline in union membership and power was also a result of the demise of manufacturing and the general state of the labour market. Brittan argues, therefore, that there was a profound change in 'atmosphere' but, by implication, not substance during the Thatcher years. He postulates, however, that Mrs Thatcher was able to popularise sentiments that had been espoused by previous Labour and Conservative Prime Ministers and Governments alike.

Thatcherism was also about a divided country[6]: those that were in work and those that were not; those that had and those that did not; prosperity in the South but not in other major areas of the country, especially the North; the rise of the 'underclass'; a significant, or perhaps terminal, decline in the British manufacturing base and a shift to service industries. The Thatcher years were also characterised by the sustained rise of managerialism and the ideology of business, a belief in the entrepreneurial spirit, a healthy but scarred corporate sector that had an increased self-confidence but needed to demonstrate decisive action[7], and a decline in state and trade union 'corporatism'. There is no doubt that a divided Labour Opposition determined on internecine warfare for much of the period up to 1988 helped Margaret Thatcher in her 'crusade'. The Thatcher 1980s also witnessed infighting within the Cabinet, high-profile leaks from the Civil Service and the loss of numerous Ministers, often in very public and damaging circumstances to the Government. One such loss, Michael Heseltine, eventually resulted in the demise of Margaret Thatcher.

This sets the broader economic and political context within which the *Faster Building for Commerce* report was produced.

10.2.2   *Industry context – the 1978 and 1983 NEDO reports*

The 1978 NEDO report *How Flexible is Construction*, is a harbinger of the two 1980s reports *Faster Building for Industry* and *Faster Building for Commerce*. It concludes that customer demand, industry capacity, process and confidence must be in harmony and be able to adapt to each other flexibly, if they cannot there will be significant costs to the industry, the economy and the country. The 1983 *Faster Building for Industry* report was given impetus by Michael Heseltine, the then Minster of State for the Environment. The subsequent 1988 *Faster Building for Commerce* report builds on the findings of the 1983 report, and the key findings from the report are reiterated here (see Chapter 8 also):

- The idea of an accelerated construction programme was unfounded and fast building could be achieved without sacrificing quality and running costs. The key role of the customer, including clear definitions of roles and responsibilities, was identified as necessary. This was equally true of the construction team.
- Projects that went well benefited from experienced customers, who supplied substantial and well-directed management input.

- Inexperienced customers were dismayed at the complexity of the traditional process. They needed advice and the industry had not made it easy for them to meet their needs. The report talks about a 'principal adviser'.
- Traditional methods of design and tendering can give good results but non-traditional methods or traditional methods overlaid with negotiation tend to be quicker.
- Prior investigation of ground conditions is essential.
- Progressing the design must be linked with facilitating progress on site and buildability. Advanced ordering of special materials should be planned, as should the input from specialist design from subcontractors.
- The contractor should be chosen on the basis not only of price but also of management capabilities. The report acknowledges the problems associated with the extensive use of subcontracting. The benefits derived from good site management are acknowledged.
- Variations may be necessary but when introduced by the design team or customer they can cause delay. Control of variations is essential.
- The attitudes of the parties are important in determining if targets are met. Standard forms of contract identify penalties for delay but not incentives for speed, and the industry and customer need to explore ways of sharing benefits from improved performance.
- Statutory agencies were identified as causing delays.

10.3   The *Faster Building for Commerce* report

10.3.1   *Aims, objectives and methodology*

The aim of *Faster Building for Commerce* was to build on the earlier research into the speed of industrial building. It formed part of an ongoing programme of work by the Building Economic Development Council (EDC) to improve construction industry performance. The agreed research objectives for the project were[1, p. 37]

> 'to explore certain issues in the management of the construction process which are central to reducing the time taken to complete building projects – in particular, the organisation of projects, the coordination of the design, the integration of subcontractors, and the management on site – to research these in complex and demanding situations in which their influence is enhanced, and to produce well-documented insights into those relationships, procedures and aspects which hold the key to improving the pace of construction.'

Detailed analysis was undertaken of a representative sample of 60 commercial projects built between 1984 and 1986. Interviews, documentation analysis and observation were utilised, with fieldwork stretching over a 2 year period. Purpose-built and speculative projects were included, as were new work and refurbishment. Projects ranged from £250 000 to £30 million, covering office and shopping developments, and those for specialist use such as hotels, hospitals and training and leisure centres. Complex medium and large projects were deliberately chosen since they were likely to challenge the 'ordinary' range of industry experience and competency. Statistical analysis was also undertaken of 8000 commercial projects, and a questionnaire was forwarded to customers of a further 260 projects. Special in-depth investigations of selected aspects of the project process were also undertaken.

10.3.2   *The structure of the report*

The report is structured along similar lines to *Faster Building for Industry*, namely:

- Part 1 – the report of the steering group;
- Part 2 – the research report;
- Part 3 – case studies;
- Part 4 – appendices.

The *Faster Building for Commerce* report is 150 pages long, including title page, foreword and appendices. It also contains two dials for assessing procurement times.

10.3.3   *Key themes*

A number of significant themes and trends are highlighted in the report and will be addressed under the following headings:

- project times;
- industry structure and process;
- the roles of key participants throughout the project.

The recommendations follow in the next section.

Project times

The research adopted two measures for assessing 'time' for projects:

- 'Total time', the period from the commencement of the project by the client to practical completion on site. The report highlights that the precontract stage has many ambiguities present and it is often difficult to determine exactly when a project commences. The start time defined by the research team is when the client commits significant resources to the project.

- 'Construction time' is easier to determine, is often stated in contract documents and is defined as the contractor's start on site through to practical completion.

The research team compared projects of similar types to determine the characteristics of fast or slow projects. Of the projects investigated, only one-third were completed on time, one-third were completed within 1 month of the time specified and the remaining third overran in excess of 1 month. One in two projects experienced delays due to unforeseen ground conditions. Each project had a unique optimum time that was determined by the customer and the building itself. However, key influences on speed were identified as:

- The extent of customer participation.
- The extent to which the customer had imposed deadlines and a rigid timescale. This was seen as transmitting a sense of urgency to all project participants.
- The quality of the design and design information.
- Control over and minimal design variations.
- The main contractor's control over site operations.
- The extent of integration of specialist subcontractors with design responsibility into the design process.

The research identified that customers often had widely differing views and expectations on preconstruction and construction times, the report proposing that increasing the awareness of normal and fast project times for differing types of project could inform debate and prevent ill-considered moves in this domain. Purpose-built and retail projects were completed faster, whereas speculative built projects were often the slowest in terms of timescales for completion, mainly owing to adjustments needed to accommodate tenant requirements.

Industry structure and process

One of the key findings to emerge from the research, documented and emphasised extensively in the report at various stages, is the fragmentation that was inherent in the industry when the research was undertaken. The findings acknowledged the increasing reliance on specialists and a myriad of autonomous organisations that were involved in delivering complex, commercial buildings. The level of fragmentation was a direct result of the sophistication and complexity of technology used in commercial buildings, the vagaries and variability of demand and the consequent increasing trend towards specialisation, subcontracting and self-employment. To overcome the increasing fragmentation of the industry and the building process, the report advocated the use of state-of-the-art management practices and systems and procurement routes that were attuned to better integration of the process at all stages. The report argues that good practices for fast, successful projects are ones where:

- There is a clear project strategy worked out and in place from the start.
- A project organisation is set up from the outset that promotes unanimity of purpose.
- Competent management exists at all stages.
- Activities and working practices are designed to interlock and leave no gaps, and roles and responsibilities are clearly understood by all.
- Time is specifically set aside to stimulate innovation and developments in the construction process.
- Project teams are so used to working with each other that they are familiar with each other's expectations and working practices.

The report is also clear that slower, poorer performing projects have the obverse of these characteristics.

A further important finding to emerge from the research is that, at the time, the procurement of commercial buildings was dominated by the use of design separated from construction – the traditional process of delivery, and using competitive tendering procedures. Over two-thirds of the projects investigated were organised along traditional lines. Design and build was the second most common form of procurement. However, the report acknowledges that the traditional method of procurement is no longer up to the task of delivering complex commercial projects, where clients will often expect to change their minds on what to include in the building, even during construction on site, in order to adjust to their own competitive pressures. Successful project delivery was characterised by a strong, high-profile role taken by clients in the management of the project under the traditional system. The research team labelled this as the 'customer-led traditional system', which differed significantly from the traditional system that had developed in the post-war period, and also from that where the architect led the project on behalf of the client.

Design and the management of the design phase had an important influence on speed. Comments were made on the adequacy of the briefing process. The continuity of design development was severely impacted by the sequence in which design and construction teams were appointed, with a consequent knock-on effect for buildability. The report clearly identified a 'hiatus' in the design process owing to detailed design work usually devolved to specialist designers. The impact of this hiatus was highlighted as being particularly acute in the design and coordination of building services. Good practice was identified as:

- Coherent management of the design phase.
- Consistency of data.
- An organised and managed flow of design information.
- The use of advanced computer facilities and information technology.
- A clear understanding between the participants over design coordination. Most customers expected the architect to coordinate the design holistically. However, the report identifies that changes in the Royal Institute of

British Architects (RIBA) terms of appointment for architects at the time no longer placed this as a duty on the architect. Many project participants were unaware of this change.

The adequacy of tender documentation was highlighted for particular criticism by the report. The contract conditions were seen as no longer able to handle the complexities of the project process, with the main contract having inconsistencies inherent within it that revolved around the increased involvement of specialist subcontractors in design and in undertaking the majority of work on site. This placed increasing importance on the conditions of contract for subcontractors, who were often appointed under different contract conditions to the main contractor. The report also highlighted deficiencies in the production of tender documentation for competitive selection of contractors, with a substantial amount of projects having insufficient information for the main contractor to price the work with certainty. Wide variations in price were identified for particular projects. The consequence was stated in unequivocal terms. Main contractors, subcontractors and suppliers would seek to restore margins in situations where the cost of work often exceeded the lowest price tendered. Equally, under a competitive tendering situation, the tender documents were the only available source of information for main contractors to understand customer requirements. The report argued that the whole area was a complex one that required resolution, understanding and acceptance and was fundamental to many of the issues highlighted in the research. The report recommended that, where information was complete, precise and certain competitive tendering was acceptable. However, where information was less definite or where major variations were expected, then two-stage tendering, negotiation or other non-traditional forms of contract should be used.

The on-site construction process was highlighted as an area of fundamental change, with the role of the site manager identified as critical to project success. The report highlighted the fact that fragmentation was particularly acute on site, with a myriad of different organisations, contract terms, employers and payment and working conditions in place. The role of the self-employed operative was singled out as having a major impact on site operations, with potentially 50% of the operatives on any site being self-employed. There was a consequent loss of control over skills and training. The site manager's job was seen as requiring a high degree of managerial and technical professionalism, but many site managers lacked any formal management training or did not have the capability to address the increasing technical specialism and sophistication of commercial projects.

The report questioned the extensive use of labour-intensive methods on site, arguing for greater use of off-site fabrication and on-site mechanisation. The report also highlighted that control over the supply, training and skills of the workforce was diffuse. There was widespread dissatisfaction with the operatives' basic rates of pay and the wage structure in general. Furthermore, bonuses and overtime were seen as the only method to make up pay, and directly employed operatives were highly critical of site management's arbitrary methods in determining bonuses. The report highlighted that site management generally failed to communicate specific information on end dates for projects or programme revisions. Operatives were highly short-term task focused. A divergence of views existed between directly employed operatives on whether they should finish on time for the reputation of the company, the minority, or would prefer to see the project take longer for continued employment.

Concerns were also expressed in the report over the supply and quality of materials, with the report urging the supply industries to move towards increased standardisation and consider using US-style catalogues of standard products as part of the drive to improve efficiency. Utilities for the supply of water, gas and electricity also raised concerns, as did statutory authorities. In the case of the former there was a lack of redress if they failed to perform, and in the case of the latter the inconsistency of decision-making between different actors in the statutory authority processes within the same authority was also highlighted. The report also raised the issue of 'planning gain', highlighting that developers were often asked to contribute something totally unrelated to the project. The report concluded that significant delay and frustration were commonplace with utilities and took a significant amount of management time to resolve to the satisfaction of the project.

The whole issue of liability and insurance was addressed in the report in some detail and was seen as a major concern to the industry. The report highlighted the increasing concern of consultants over the cost of professional indemnity insurance. The report acknowledged that the increasing complexity of commercial buildings and the diffuse responsibility for design increased the difficulty of project coordination. Determining liability in the event of things going awry had become more difficult, and the report highlighted that parties to the project had become much more focused on manoeuvring to reduce exposure to liability for their work. Equally, customers were more willing to seek legal redress, and the report proposed that this whole area had added to the adversarial climate of the industry. In order to improve the efficiency of the process and reduce the requirement to determine fault, the report supported the development of materials damage insurance.

The image of the industry was perceived as not offering a desirable or rewarding career in the UK. It was proposed that the industry should not only address this aggressively at all levels but also attempt to widen the attractiveness of the industry to ensure a ready supply of capable and well-trained candidates. The report argued that universities and polytechnics should widen entry requirements to facilitate arts-based candidates entering the industry. No full-time academic was on the steering group to provide inputs into that debate.

The roles of key participants throughout the project

This subsection explores the impact of changes to the different roles identified in the report. Key roles to be explored are:

- the customer,
- designers,
- the main contractor and site management,
- subcontractors,
- operatives.

The report concluded that customers were a key influence on the progress and outcomes of projects. Regular procurers of commercial buildings appreciated the complexity and difficulties inherent in the process and received a better level of service from the industry owing to the potential for repeat orders, with many having standardised procedures for dealing with the industry. They considered in-house executive procurement knowledge to be essential and often had networks of regularly used consultants, contractors and suppliers from which teams used to working together would be assembled to ensure certainty of completing on time and that the achievement of other project objectives was regularly attained. In this sense the report was a precursor of the partnering concept which emerged some 15 years later. The report argued that the inexperienced customer had great difficulty in getting good, impartial advice about appointing teams. It was proposed that a role of 'customer representative' be created with no vested interest in any procurement route and with no other function. Design was no longer seen as the sole domain of consultants, with much of the detailed design work undertaken by specialists subcontracted to the main contractor. Specialists were seen as having a pervasive impact on the success of projects.

The report acknowledged a dramatic change in the role of the main contractor. A significant amount of the work on site was highlighted as being undertaken by different types of subcontractor. The main contractor's role was now one of organising and coordinating others and of procuring materials and other resources. Their expertise focused on management and their ability to carry and manage contractual risk. The report acknowledged that they were increasingly being drawn into accepting contractual responsibility for subcontractor design inputs, and this element of their role required clarification and formalisation. As already established, the site manager role was seen as critical, and, again, changes were in evidence for its requirements, with the role now comprising coordination, progress chasing and acting as a chaser of inputs from others. The role of the main contractor was now seen as differing little from that of the management contractor. The high incidence of labour-only subcontractors created tensions within site operations. The report highlighted that directly employed operatives felt increasingly marginalised and were perceived to be there to support the self-employed operatives. Directly employed operatives were also highly critical of labour-only subcontractors (LOSCs) in terms of safety and regard for the orderliness and cleanliness of the site and were blamed for the fragmentation of the workforce. Conversely, LOSCs, generally younger than directly employed operatives, took the view that they were unable to be employed as directly employed workers. However, they considered themselves capable of high output and expected to be attended by main contractor's operatives.

The report did, however, also highlight operatives' pride in their work, particularly among the more highly skilled craftsman, who also had views on design and buildability. Operatives welcomed the opportunity to demonstrate such knowledge and skill. LOSCs preferred to work on designs that were more straightforward and provided opportunities for high output. Where the roles of site management and operatives produced a climate of 'them and us', a perception existed in the former that the latter were only concerned about pay. On such sites, operatives felt that management was remote and consequently treated them with suspicion and distrust. On sites where the management had created a climate of wider relationships with operatives, pay was not such a dominating feature and a climate of trust was engendered.

This brief review of key themes does not cover all of the issues raised in the report. Consequently, summaries of the major conclusions will be presented.

10.3.4   *The report's recommendations*

Twenty recommendations were produced by the steering group:

1 The range of services provided by members of professional and trade bodies should include that of independent customer representatives. The bodies should advertise the availability of the service and provide lists of suitable people.
2 Designers should make more use of models as a means of communicating design concepts.
3 Conditions of appointment for professionals should be harmonised to facilitate coordination of design.
4 Achievement of material damage insurance, covering building elements such as structure and weathershield, is vital.
5 Procedures such as early appointment should be used more often to enable contractors to advise on buildability.
6 Designers' training should emphasise the importance of buildability through a period of practical site experience.
7 The method of selecting the main contractor must fit the circumstances of each individual contract.
8 Clarification is required of the role of the main contractor in relation to design inputs from subcontractors.

9 The terms of appointment for subcontractors should be reviewed.
10 Allocation of responsibilities between main contractor and subcontractors should be clear.
11 A rethink of the relationship between management attitudes and the standard forms of contract is required with a view to initiating fundamental changes to reduce conflict.
12 Entry requirements and curricula of courses in universities and polytechnics should be reviewed in relation to modern management requirements.
13 Contractors should involve themselves further with academic institutions and sponsor students.
14 A study of productivity should be mounted.
15 The grading and pay structure of the operatives should be rationalised.
16 US-style catalogues of standard products should be developed.
17 The industry should mount a major PR campaign to promote its successes and career prospects.
18 A single officer should coordinate all decisions within a local authority in relation to large projects.
19 The statutory basis of Section 52 agreements and similar arrangements should be revised to make them simpler and more certain in operation.
20 Public utilities should provide:

   (a) easy access to their services;
   (b) design and advisory services;
   (c) detailed cost estimates;
   (d) supervision of subcontractors;
   (e) binding contracts;
   (f) an efficient and responsive complaints procedure.

Each of these recommendations proposes organisations to initiate action. The next section provides a commentary on the report.

10.4 A commentary

The foreword for the report by Sir Christopher Foster, Chairman of the Construction Industry Sector Group, identifies that, at the time of writing, the UK was at the centre of commercial development in Europe. He acknowledges that the industry had made trail-blazing improvements during the 1980s but those needed to become the norm. He explicitly stated that, if that came to pass, customers would be satisfied, and the industry would be more able to compete overseas as well as against foreign competitors in the UK home market. Certainly, in the 1980s the UK witnessed a significant intrusion of foreign competition into its domestic markets, mainly through acquisition. There is a sense that the 1983 *Faster Building for Industry* and 1988 *Faster*

Building for Commerce reports were preparing the industry to meet increasing global competition and ensure that the best of the UK firms were able to compete internationally by getting closer to customers, upgrading management skills and tailoring flexible approaches in procurement to assist different customer requirements. Equally, both reports allude to increased customer care in the domestic market to ensure that international competitors meet tough domestic competition in the UK.

The *Faster Building for Commerce* report, does, however, produce a gloomy picture of the industry, one that has lost control or is certainly in danger of losing control over the process: on site, where tensions exist between different types of operative; site management not being up to the job; forms of contract that are inappropriate for the new roles emerging in the industry; and designers that are no longer capable of managing the design process without strong, influential inputs from the client or being aware of the impact of buildability of their design. The report adds nothing to improving the image of the industry. From a rational perspective, however, an interpretation of the report suggests that contractors have responded to the dynamics of the industry they face – adopting management strategies that permit them to adjust flexibly to the business environment by outsourcing manpower. A perverse view, given the dominance of the traditional system of procurement using competitive tendering, endemic in the projects studied by the research team, is that clients and their advisors, including Government, have got the industry they have created. Demand is a client and Government policy driven phenomenon.

The rise of the self-employed subcontractor, demonstrated admirably in the report with some force, could be interpreted as a natural consequence of the entrepreneurial spirit espoused by Margaret Thatcher. What is clear, however, is that it shaped the industry then and continues to shape the industry today. As such, the *Faster Building for Commerce* report highlighted a fundamental shift in the management of site operations and defined or at least redefined new roles in the industry (19 pages are devoted in the research report to managing these changes during the construction phase). The new role of the main contractor set out in the report is the forerunner of the 'prime contractors' and 'principal supply chain managers' of today.

The report also leaves one with the sense that the industry takes good care of its regular procuring clients, but ad hoc, less knowledgeable clients lose out. Perhaps the role of the independent 'client advisor' recommended in the report and the rise of consultant project managers will meet their needs. It is unlikely. The whole area of the adequacy of the service that the industry provides to ad hoc, irregular procuring clients needs in-depth analysis and contrasted inadequately with that provided to regular volume procurers.

There is no doubt that the *Faster Building for Industry* and *Faster Building for Commerce* reports in combination heralded the arrival of attuning procurement routes to the needs of clients and projects, ensuring strategic fit. Also, around that time value management and value engineering were appearing on the scene; architects were retreating into offering a design service at the expense of a broader involvement in the process; quantity surveyors were redefining their role yet again; briefing was again singled out as a key process to engender a dialogue between 'customers' (or clients) and their delivery teams to ensure the latter understood the former's requirements. The *Faster Building for Commerce* report highlighted critical elements for project success that remain today: effective team working; teams that work together regularly; good 'clienting' practice, especially surrounding briefing; a thorough understanding by those involved in the process of the expectations, roles and responsibilities of the contributors to the process, updating contractual terms and conditions to reflect the complexity in the design and construction process. The roots of the current trends towards partnering, supply chain management, continuous improvement and bringing the right team together at the right time are clearly indicated in the *Faster Building for Commerce* report. So what has changed since then?

10.5   Conclusions

The report *Faster Building for Commerce* was initiated in 1984. It took 4 years to be researched and published. Its predecessor, *Faster Building for Industry*, adopted much the same methodology and took approximately the same time to be researched and published. The interlinked reports took 8 years of resource commitment to come to fruition. Unfortunately, the former is not presented as well as the latter. However, in combination, the two reports provide a comprehensive analysis of, and document changes in, the industry. They raise issues and make recommendations about the organisation, procurement and management of projects. They discuss good and inappropriate practice. Why were the Latham and Egan reports necessary? There has been a peculiar tendency in the industry to self-deprecate regularly and over a sustained period of time, from the immediate post-war period onwards. Is it any wonder that the industry does not have a good image and is not able to recruit the best candidates?

The genesis of the Latham report occurred within only a few years of the *Faster Building for Commerce* report being published. Egan followed on the heels of Latham. The latter certainly created a range of initiatives, and Egan much the same. However, similar general themes re-emerged in the 1990s when Latham and Egan were published, except that the latter also drew on insights from manufacturing. Perhaps the industry is now ready to adapt and change much faster and relinquish its embedded ways of adversarial contracting. However, change takes time to bed in and to be reflected on and assimilated. Trust goes beyond contracts, it is deeply rooted in interpersonal behaviour. Contracts exist primarily to protect self-interest and avert overt opportunistic behaviour in a commercial world. It is difficult to formalise and 'contractualise' trust, and the jury is out on how effective standard partnering contracts and other types of collaborative procurement route will be in improving the industry's adversarial history, particularly if underlying inter-personal behaviour does not change.

10.6   References

1   NEDO (1988) *Faster Building For Commerce*. National Economic Development Office, HMSO, London.
2   NEDO (1978) *How Flexible is Construction?* National Economic Development Office, HMSO, London.
3   NEDO (1983) *Faster Building for Industry*. National Economic Development Office, HMSO, London.
4   Riddell, P. (1988) An instinct not an ideology. In: *The Thatcher Years: The Politics and Prospects*, pp. 5–7. Financial Times.
5   Brittan, L. (1988) *The economy: traumatic rather than radical*. In: *The Thatcher Years: The Politics and Prospects*, pp. 13–17. Financial Times.
6   Rogaly, J. (1988) Divided Britain. In: *The Thatcher Years: The Politics and Prospects*, pp. 42–6. Financial Times.
7   Lambert, R. (1988) The corporate sector: healthier but the scars still show. In: *The Thatcher Years: The Politics and Prospects*, pp. 46–8. Financial Times.

Chapter 11

Constructing the Team:
The Latham Report (1994)

Daniel Cahill & Marie-Cécile Puybaraud

11.1 Background to the study: the economic, political and social climate

According to the 1994 Latham report:

> 'The recession of recent years has hit the construction industry very hard, though hopefully some improvement in trading conditions is now beginning. It affected the construction industry more deeply than other industries.'

The 1980s saw considerable changes in economic, political, social and technological activity. The emphasis on the service sector and a decline in the manufacturing sector focused much of the economic growth experienced to the South-East of the UK. The construction industry benefited considerably from this economic growth and grew in size and capacity accordingly. This period of exceptional growth came to an abrupt end with a very deep recession that started in 1989/90.

Hillebrandt et al.[1] outlined the interaction of the construction industry and the general economic conditions through the period immediately prior to the Latham review. They argued that the sudden tightening of monetary policy in 1988 and its impact on the housing and property markets were the trigger for the recession that afflicted the industry throughout this period. This in turn played a significant role in the recession in the general economy, and the collapse of the housing market, the rapid rise in interest rates and falling house prices all affected consumer confidence. This decline in confidence only served to compound the problem by leading to significant reductions in sales of consumer goods which in turn affected the retail and manufacturing sectors, leading to widespread bankruptcies and lending further fuel to the economic turmoil of this period.

At the height of this recession, a 'bid low, claim high' approach tended to prevail, whereby consultants and contractors sought to obtain work at all costs, often at a loss, in the hope that an element of profit could be recouped through claims and extras as the project progressed. This practice obviously created an increasingly adversarial and conflict-driven business environment, and there was growing dissatisfaction with the industry by many parties, particularly the Government.

The main difference of the Government response to the effects of this recession compared with earlier ones was not immediately to try to stimulate the economy through significant investment in buildings and infrastructure, but to leave the economy to the mercies of private investment and market forces because of the general policy of reducing the role of central and local government through privatisation of many of the services it originally provided.

11.2   The *Constructing the Team* report

11.2.1   *Introduction*

The *Constructing the Team* report published in 1994, more generally known as the Latham report, was a watershed document for the construction industry. Much of its thrust was for a more cooperative, less adversarial construction industry. Latham argued that a healthier atmosphere was the key to enhanced performance, and partnering was identified as a means to this end. The endless refining of contract conditions would not solve adversarial problems, so a fresh approach was required.

Latham carried out his review at a very difficult time for the industry: it had just experienced the heights of economic growth a few years earlier, only then to 'hit the buffers', and it was currently experiencing significant decline and was in a very poor state as noted above.

Prior to the publication of Latham's report, an interim report, *Trust and Money*, was issued in December 1993. This latter concentrated on an analysis of the main problems that those consulted had outlined to the review team, as well as reiterating some long-standing best practice advice.

11.2.2   *The work of Latham*

Latham described his report as follows:

> 'It is a personal report of an independent, but friendly, observer. No blame attaches to anyone except myself for its content.'

The personal views of Latham were strongly put across in the report, and Latham acknowledged full responsibility for any outcomes and recommendations in it. Barnicoat[2] argued that 'Sir Michael's enthusiasm and commitment has not waned. He is still pushing for change in the industry'.

As well as funding from the Department of the Environment, Latham received the support of four industry funding organisations:

- Construction Industry Council;
- Construction Industry Employers Council;
- Specialist Engineering Contractors Group;
- Nationalist Specialist Contractors Council.

and two client groups:

- Chartered Institute for Purchasing and Supply;
- British Property Federation.

They were not intended to act as a decision-making committee or drafting body. This was very much an individual review carried out by Latham. The funding committee supported him by acting as an expert sounding board and conduit of advice for his review.

11.2.3   The recommendations

In summing up his report, Sir Michael Latham identified the 30 key recommendations presented in Table 11.1. The recommendations were intended to be taken as a package. Individual organisations involved in the construction process welcomed some of the recommendations while finding others more difficult to accept. Industry opinion seemed to be very positive, an attitude survey conducted for *New Builder* magazine[3] reported that, out of the ten main proposals, senior industry executives

> 'gave an unambiguous thumbs up to four, cautious support to another five and only rejected one outright. Across all ten proposals, the contractors were the most positive, followed by the housebuilders, the consultants, the clients and the material producers.'

However, overall, the implementation of the report meant that every sector would gain through the introduction of more efficient business practices. It would provide a better definition of needs and objectives and of roles and responsibilities. The end result would be a clearer framework and a more stable environment for the industry.

Key among the recommendations was 'a productivity target of 30% reduction in real construction cost reduction by the year 2000'. The target was considered as the motivating factor for the implementation of all the other proposals in the report.

Other important recommendations included:

Table 11.1 The Latham report recommendations.

Recommendation	Notes
1 Focal Points – clients	The DOE (Department of the Environment) should act closely with the Departments for Scotland, Wales and Northern Ireland.
	Government should commit itself to being a best practice client.
	A construction forum should be created to represent private sector clients.
2 Guide for clients on briefing	The CIC (Construction Industry Council) should prepare a guide to briefing (check list). It should also be part of the contractual process that the client should approve the design brief by 'signing it off'.
3 Code of practice	The DOE should coordinate and publish a construction strategy code of practice (CSCP) to inform and advise clients.
4 Code of practice	The CSCP should be designed to assist clients to meet their objectives and to obtain value for money. The guide should also be designed to harness clients' purchasing power to improve the long-term performance of the industry.
5 Consultation of the Process Plant Sector	The European Construction Institute should be involved in the implementation of the CSCP.
6 Check-list designers	The formulation of a check list, or the adoption of existing ones such as those of the BPF (British Property Foundation) or BSRIA (Building services), should be an urgent task of the reconstituted JCT as part of their new duties.
7 Coordinated project information (CPI)	The CPI technique should be made part of the conditions of engagement of the designers and be made a contractual requirement.
8 Allocation of M&E (Mechanical and Electrical) design responsibilities	Whatever procurement system a client employs, the allocation of design responsibilities between design consulting engineer and specialist engineering contractor should follow the check list of guidance with a separate design agreement for the specialist engineering contractor.
9 The Joint Contracts Tribunal	To amend the standard JCT and ICE forms to take account of the principles set in the report.
10 The CCSJC (Construction Contract Standards Joint Committee)	The structures of the CCSJC and the JCT need substantial change.
11 Joint Liaison Committee	A joint liaison committee should be formed to consider amendments to the NEC and to build up a complete family of documents around it.
12 Clients	Public and private sector should begin to use the NEC, and phase out 'bespoke' documents.
13 Register for consultants and quality/price assessment	There should be a register of consultants kept by the DOE for public sector work. Firms wishing to undertake public sector work should be on it.
14 Project sponsors and managers	The roles and duties of the project sponsor need clearer definition. Government project sponsors should have sufficient expertise to fulfil their roles effectively.
15 Main contractors' and subcontractor list	A list of contractors and subcontractors seeking public sector work should be maintained by the DOE. It should develop into a quality register of approved firms.

Table 11.1 Cont.

Recommendation	Notes
16 Tendering	Detailed advice should be included in the CSCP to all public sector clients on the specific requirements for selective tendering of European Union Directives. Advice should be issued on Partnering arrangements.
17 Interim arrangements	The DOE should set up a central qualification list based on subcontractors and contractors seeking public sector work. Such a list should be supported by a national scheme guidance for quality assessment of tenders.
18 Selection of subcontractors	A joint code of practice for the selection of subcontractors should be drawn up which should include commitments to shorter tender lists, fair tendering procedures and teamwork on site.
19 Partnering	Advice should be given to public authorities on Partnering so that they can experiment with partnering arrangements where appropriate long-term relationships can be built up. Any partnering arrangements should include mutually agreed and measurable targets for productivity improvements.
20 Training	Recent proposals relating to the work of the Construction Industry Training Board (CITB) need urgent examination.
21 Image of the industry and equal opportunities	The industry should implement recommendations that it previously formulated to improve its public image. Equal opportunities in the industry also require urgent attention, with encouragement of the Government.
22 Professional education	The CIC is best suited to coordinate the implementation of already published recommendations on professional education.
23 Research & development	Existing research initiatives should be coordinated and should involve clients. A new research and information initiative should be launched, funded by a levy on insurance premia for mandatory 'BUILD' type insurance.
24 Productivity target	A productivity target of 30% real cost reduction by the year 2000 should be launched.
25 Unfair conditions	A Construction Contracts Bill should be introduced to give statutory backing to the newly amended standard forms, including the NEC. Some specific unfair contract clauses should be outlawed.
26 Adjudication	A system of adjudication should be introduced within all the standard forms of contract, and this should be underpinned by legislation. Adjudication should be the normal method of dispute resolution.
27 Trust funds	Mandatory trust funds for payments should be established for construction work governed by formal conditions of contract.
28 Liability legislation	The Construction Contracts Bill should implement the majority recommendations of the working party on construction liability law.
29 Latent defects insurance	The Construction Contracts Bill should contain a provision for compulsory latent defects insurance for 10 years from practical completion for all future new commercial, retail and industrial building work.
30 Possible delivery mechanisms	An implementation forum should monitor progress and should consider whether a new Development Agency should be created to drive productivity improvements and encourage teamwork.

- the clear establishment of *Government as a best practice client* which would set an example for other clients to follow;
- the production of a *client's guide* to briefing to assist all clients in understanding and being involved in the drawing up of briefs for projects;
- the preparation of a construction strategy *code of practice* to provide a guidance on best practice, including contract strategy, tendering and project management;
- the introduction of a complete *standard family of contract documentation* enshrining the basic principles set out in the report;
- the *rationalisation of the systems* by which construction companies were asked to prequalify for contracts;
- the development of *mechanisms for the selection of consultants* which would allow both price and quality to be given appropriate consideration.

The report, which was widely welcomed, set a series of objectives with very ambitious timescales to tackle problems and to lay the foundations for a more successful, less confrontational industry in the future. Most of these objectives were set out in an action plan with ambitious timescales, and these are critically examined in Table 11.2 in terms of the desired objectives and the real outcomes.

11.3   Post-Latham developments

Sir Michael Latham's proposals were warmly supported by all political parties throughout the country:

'One year has passed since the publication of the Latham Committee report and, to the Government's credit, it has not been pigeon-holed and forgotten.'

Commons Hansard (19 July 1995), House of Commons Hansard Debates for 19 July 1995

Members of Parliament and the representatives of the construction industry and companies in our constituencies want to see such legislation introduced.
Latham's recommendations cover three major aspects of the construction process:

- modifying and adapting the *legislation* to address the wider need of the industry;
- reviewing construction *contracts* to meet Latham's principles;
- addressing the *procurement* to integrate the strategic business model into the construction process.

Table 11.2 Latham's action plan.

Recommendations for actions	Latham's target	Action for implementation	Comments
Government as best practice client	Sept. 1994	1995 Construction Industry Board	The CIB was set up to improve the performance of the UK construction industry. The Board is the forum that brings together suppliers and customers from private and public sectors with central government.
		1995 Levene efficiency scrutiny into construction procurement by Government	HM Treasury: procurement guidances 1 to 9 between 1997 and 1999. Publication of a series of Government construction procurement guidances. This series of guidances is the result of the 1994 and 1995 reviews, and particularly the efficiency scrutiny. The guidance provides best practice advice at a strategic level.
			The CIB was closed down in June 2001, but many of its key projects are continuing independently until completed or fall under the remit of the Construction Industry Council.
Creation of Construction Client's Forum	End 1994	Construction Client's Forum, launched in 1998 by the CIB	The Forum represents the client on the CIB.
Guide on briefing	End 1994	Central Unit on Procurement	CUP Guidance 1995–1997
Construction strategy code of practice	July 1995	Central Unit on Procurement	CUP Guidance 1995–1997
Joint of code of practice for selection of subcontractors	End 1995	CIB	Publication of a joint code of practice for the selection of subcontractors, Working Group 3, April 1997.
Restructuring contracts	End 1994 Phasing out: 1994/ 1998	Major review and republication of all JCTs and GC (Government Contracts) contracts in 1998	JCT amended and republished a new set of contracts in 1998 taking into account Latham's recommendations and legislative changes.
			GC published a new edition of the standard Government forms of contract in 1998, following the recommendations of Latham, the HMSO report *Construction Procurement in Government* in 1995 and the implementation of the Construction Act 1996.

Cont.

Table 11.2 Cont.

Recommendations for actions	Latham's target	Action for implementation	Comments
Use of NEC	ASAP	March 1993: first edition of NEC published	Limited use of NEC, *Building* (1999), indicated an average of 5% on Government funded projects.
		1994: first edition of Adjudicator's Contract and Professional Services Contract	
		July 1994: The Latham Report is published and endorses NEC	
		1995: NEC second edition published as the NEC Engineering and Construction Contract (ECC)	
		November 1995: NEC second edition reprinted with minor amendments	
		April 1988: Amendment published to comply with UK Housing Grants, Construction and Regeneration Act (1996)	
		June 1998: second edition of Adjudicator's Contract and Professional Services Contract	
		May 1998: NEC second edition reprinted with corrections	
		July 1999: Engineering and Construction Short Contract published	
		October 1999: Chinese translation of the Engineering and Construction Contract with guidance notes published	
		April 2000: Amendment published for Contracts (Rights of Third Parties) Act 1999 – Y(UK) 3	
		2000: NEC Partnering Agreement issued for consultation	

Table 11.2 Cont.

Recommendations for actions	Latham's target	Action for implementation	Comments
Working party	End 1994	The Client's Forum The Housing Forum	Taken over by Egan (1998)
Expansion of ConReg	July 1995	Government Construction Act 1996	The Housing Grants, Construction and Regeneration Act 1996, designed to address issues of late payments and dispute resolution came into effect in May 1998. The Act deals with the pay-when-paid clauses in contracts, and contracts must now have a mechanism for determining the amount and timing of payments, as well as the right for adjudication of disputes.
Examination of JAGNET (Joint Action Group Network Engineering Training) report and CIEC proposals	July 1995	Government/CIB	
Task-force on the image of the industry	October 1994	Action taken over by the 1998 Egan report	M4I: Movement for Innovation/Respect for People
Equal opportunities	July 1995	Action taken over by the 1998 Egan report	M4I: Movement for Innovation/Respect for People
Professional education task force	July 1995	CIB, CITB, DTI, CBPP Forums, Working groups	Construction Best Practice Programme (CBPP) The CBPP has teamed up with the Construction Industry Training Board (CITB) to develop an important series of workshops – called Learning by Doing.
Liability proposals BUILD insurance Research levy	Brief formulation by end 1994	Construction Act 1996	Amendments integrated into all standard forms of contracts and republished in 1998.
Productivity initiative	By End of 1999	1995 Construction Industry Board	Taken over by Egan (1998)
Development Agency	April 1995	Construction Best Practice Programme (CBPP)	

11.3.1   Legislation

One significant piece of legislation resulted from the Latham report, namely the Housing Grants, Construction and Regeneration Act 1996 (Construction Act). This prompted major amendments to standard form construction contracts.

The Housing Grants, Construction and Regeneration Act 1996 was designed to address issues of late payments and dispute resolution came into effect in May 1998. The Act deals with the pay-when-paid clauses in contracts, and contracts must now have a mechanism for determining the amount and timing of payments, as well as the right for a rapid and impartial adjudication process to resolve disputes.

Murdoch & Hughes[4] set out the changes brought about by this Act, in particular Part II, which lays down rules that are applicable only to construction contracts.

The first point that is worth noting is the treatment of construction contracts as a separate class of contract. This required the specific definition of a construction contract. This is quite wide ranging and includes any agreement under which a party carries out recognisable construction work involving site clearance, the construction, alteration, repair, maintenance, extension, demolition or dismantling of buildings, or structures, including the installation of services, and any associated works, such as roads, water mains, sewers, etc. It also covers professional services by including agreements to do architectural, design or surveying work.

There are some notable exceptions such as oil or natural gas exploration work, work in the process industry, manufacture or delivery to site of typical building services components and most construction contracts with a residential occupier.

Any construction contract that complies with the definition must then satisfy the Act with regard to two specific areas, adjudication and payment. If it fails, then a scheme for construction contract applies, laying down detailed rules that are applied to ensure that it then complies with the Act.

Adjudication

The intention of this section of the Act is that the parties to a construction contract have the right to a speedy and inexpensive dispute resolution procedure that is binding on both parties but that can be challenged later at arbitration or in court. Contracts must ensure that a dispute is referred to an adjudicator and that a binding decision to resolve the dispute is completed within a very short period of time. This decision may be accepted as binding or it may be referred to arbitration or the courts at the completion of the project:

> 'A party to a construction contract has the right to refer a dispute arising under the contract for adjudication ...'.

> The Housing Grants, Construction and Regeneration Act 1996 Part II,
> Section 108 (1) Adjudication

'Let me also remind the House that the adjudicator is a creature of the contract. Interests from all sides of the construction industry have stressed throughout our proceedings on the Bill that the adjudicator should not be confused with an arbitrator, whose quasi-judicial role is spelt out at length in statute.'

The Minister for Construction, Planning and Energy Efficiency (Mr Robert B. Jones) Commons Hansard (8 July 1996), House of Commons Hansard Debates for 8 July 1996 (pt 23). New clause 2

Payment

Part II of the Housing Grants, Construction and Regeneration Act 1996 provides a framework for fairer contractual arrangements and better working relations within the construction industry:

'A party to a construction contract is entitled to payment by instalments, stage payments or other periodic payments for any work under the contract ...'.

The Housing Grants, Construction and Regeneration Act 1996 Part II, Section 109 (1)

When implemented correctly, construction contracts have to provide a rapid and impartial adjudication process and a number of rights and responsibilities relating to payment. In the event that the contract does not comply, the rules of a scheme of construction contract will be applied to the contractual situation to ensure that it does adhere to the legislation.

11.3.2   Contracts

The philosophy of contracting plays a very important role in the UK construction industry. Two major recommendations emerged from Latham's report:

- a need to review and improve standard forms of contracts in the UK;
- a need to introduce better working practices in the construction industry by adopting a partnering approach and thereby reducing conflicts and disputes.

The complexity and volume of current standard forms of contract were among the areas addressed in the Latham report. It advocated a thorough overhaul of the present standard forms to make them simpler, clearer, non-adversarial and more in tune with today's construction industry. Latham suggested using the New Engineering Contract (NEC) as the framework for reform.

The Latham report advocated the following:

1 A specific duty for all parties to deal fairly with each other.
2 Firm duties of teamwork, with shared financial motivation to pursue those objectives.
3 A wholly interrelated package of contract documents.
4 Easily comprehended language.
5 Separation of roles of contract administrator, project or lead manager and adjudicator
6 A choice of allocation of risks.
7 Taking all reasonable steps to avoid changes in preplanned works information. However, where variations do occur, they should be priced in advance, with provision for independent adjudication if agreement cannot be reached.
8 Express provision for assessing interim payments by methods other than monthly valuation.
9 Clearly setting out the period within which interim payments must be made to all participants in the process with compensation for slow payment.
10 Providing for secure trust fund routes of payment.
11 Speedy dispute resolution by a predetermined impartial adjudicator/ referee/expert.
12 Providing for incentives for exceptional performance.
13 Making provision where appropriate for advance mobilisation payments (if necessary, bonded) to contractors and subcontractors, including in respect of off-site prefabricated materials provided by part of the construction team.

The NEC claims to be suitable for most forms of procurement for both building and civil engineering work. Very briefly, it comprises a basic document of core clauses that always apply. To this the user adds a set of optional clauses specific to the procurement method used. It is clearly written in simple language and is very much shorter than existing standard forms. In his report, Latham argues that

> 'the NEC contains virtually all of these assumptions of best practice, and others, which are set out of the Core Clauses, the main and secondary options.'

Despite the initial enthusiasm for the NEC, anecdotal evidence suggests that its use has not been widespread and tends to be used mainly within the civil engineering side of the industry.
Broome[5] suggested that

> 'as a stimulus to good management the NEC has been designed as a management tool to encourage proactive participation by the parties so as not to allow the escalation of disputes and to promote a partnering approach to the whole project process.'

11.3.3 *Procurement practice*

Partnering is a modern management practice gaining increasing interest and becoming more widely practised within the UK construction industry. It is a novel supplement to traditional contracting relationships that has gained credibility from evidence of proven success in the USA, and more recently from players within the UK in other industries.

The growing popularity of partnering in the UK can be ascribed to a number of factors:

- its increasing maturity as a management approach in the USA, and recorded evidence of its success;
- the current post-Latham climate in the UK industry which has made the conditions right for partnering.

The growing emphasis is on more sophisticated management forms of procurement, used to procure ever more complex buildings. Partnering is a means of reducing the greater risks and possibility of disputes associated with these conditions.

Although partnering aims to avoid disputes, it would be ludicrous to suggest that any management philosophy can eliminate all the disputes that can stem from problems occurring on site in the course of a project. Partnering can minimise the likelihood of such disputes, but it is inevitable that in an undertaking as complex as procuring a building, with its many participants and complex contractual relationships, disputes will arise from time to time.

This recommendation has resulted in the more in-depth review of industry practices with regard to procurement, and a number of guidelines outlining good practice have been published. The Construction Industry Board (CIB)[6] and Bennett & Jayes[7,8] have developed guides encompassing Latham's recommendations.

11.4 Final thoughts

In a Commons speech in July 1996, Nick Raynsford stated:

'The Latham recommendations have in theory been endorsed by the Government, but the process of putting them into effect has been somewhat tortuous.'

Nick Raynsford, Commons Hansard (8 July 1996), House of Commons Hansard Debates for 8 July 1996 (pt 23). New clause 2

Pollington[9] argued that Latham's report 'ushered in a new era of collaboration between supply and demand sides of the construction industry':

> 'One year has passed since the publication of the Latham Committee report and, to the Government's credit, it has not been pigeon-holed and forgotten.'
>
> Commons Hansard (19 July 1995), House of Commons Hansard Debates for 19 July 1995

Latham argued that the message of his report was strongly reinforced by *Rethinking Construction* in 1998[10]. In 2000, the Government was still reviewing the future prospects of the construction industry and, in a debate at Westminster in May 2000, the Minister for Housing and Planning expressed his concerns. Nick Raynsford's introductory speech summarised the state of the UK construction industry in 2000 and highlighted latent problems into the twenty-first century:

> 'The construction industry is important for a number of reasons, one of which is its scale. It has an annual turnover of £65 billion – some 8% of gross domestic product – it employs 1.4 million people or 1.7 million if the professionals and consultants are included, and it consists of 160 000 companies. But it is not just size that matters. The construction industry is important because of its vital role as a key delivery mechanism for the improvement of our economic and social infrastructure. The public sector alone accounts for some £25 billion of the industry's turnover, which is 40% of the total. Railways and other public transport facilities, hospitals, schools, universities, housing and other public services depend on high-quality construction. Similarly, in the private sector, services and industries depend on efficient, reliable, good-value construction for the delivery of plans to improve their services to consumers.
>
> 'The Government came to power with a commitment to invest in the modernisation of Britain and to improve its economy and social infrastructure. Construction plays a vital role in the delivery of those improvements. British construction companies are world leaders in many respects. They are renowned for innovative engineering, high-quality design and architecture, and they are world leaders in project finance. They are major players in the supply of materials and components, project management and professional consultancy. They earn around £10 billion a year in overseas earnings.
>
> 'However, the industry also has problems. It has traditionally been fragmented and even the largest United Kingdom contractors are small in comparison with the major continental, American and Japanese companies and, in a globally competitive market, that sometimes matters. Profit margins have traditionally been low and construction companies have seen relatively poor stock market valuations.'
>
> The Minister for Housing and Planning (Mr Nick Raynsford), Westminster Hall, Thursday 18 May 2000, Construction Industry

Latham's targets and subsequent achievements are critically examined in Table 11.2. It would be wrong to say that nothing was achieved as suggested by quotes from senior industry executives in *Building*[11]:

- 'The report has no impact whatsoever. It merely made some people stop and think, and then they continued as before.'
- 'To date, the Latham report has had very little, if any, effect on the construction industry.'

The perception within the industry may be that very little has changed, but in reality there has been a subtle transformation of the culture of the industry. Individually, the targets were realised to some extent or other, but not necessarily within the ambitious timescales. Collectively, the recommendations did achieve the main aim of the report, which was to encourage reform to reduce conflict and litigation and encourage the industry's productivity and performance. Clients, designers and contractors have changed their way of operation in response to these actions. The increasing pressure on the construction industry to improve value for money has forced businesses to adopt new production methods, such as prefabrication, and innovative procurement methodologies, such as partnering. Another significant change as noted in *Building*[12] is that the burden of resolving construction disputes has to an extent been shifted away from the courts to adjudicators. These changes, together with the variations in business conditions over the period, have resulted in a major transformation of the industry. This is confirmed by the following from Chevin in *Property Week*[13]:

'Government-backed reforms championed in the mid-1990s by Sir Michael Latham and reinforced by Sir John Egan are now bearing fruit. Both reports advocated a move away from lowest-price tendering and a claims culture. Anecdotal evidence certainly suggests that by pricing more realistically at the outset and working more closely with clients, consultants and subcontractors, the claims have been diminishing and clients have been getting a little closer to their holy grail of getting projects completed on time and to budget. Add to this mix the influx of outsiders running companies, increasing Government pressure to improve safety, and, at the top end of the businesses at least, contractors are starting to look more professional all round. The British Property Federation's director general, Will McKee, has observed the change of heart:
'I think that Latham and Egan have changed the culture in the industry. There are fewer complaints about builders than a few years ago.'

The final word must be left to Latham. Ever the pragmatist, he is quoted in *Building*[14] as saying:

'It's true that Government has been slow to improve, and that's disappointing. I've always said the impact of my report is better than I'd expected but not as good as I'd hoped. But it is quite wrong to say there was no impact at all.'

Indeed, in a recent retrospective examination of the report's impact, Latham argues that it is wrong to conclude that the only real outcome was the Construction Act. He notes that very few of the 53 recommendations made in the report required legislation[14]. This in itself points to a report that has encouraged the industry to adopt cultural change from within, rather than through Government intervention.

11.5   References

1   Hillebrandt, P., Cannon, J. & Lansley, P. (1995) *The Construction Company in and out of Recession*. Macmillan Press, London.
2   Barnicoat, C. (1999) *Towards Zero Defects Culture*. GP Forum, No. 6.
3   *New Builder* (1994), Latham feedback. 7 October, 16.
4   Murdoch, J. & Hughes, W. (2000) *Construction Contracts, Law and Management*. 3rd edn. E & F Spon Press.
5   Broome, J. C. (1997) A comparison of the clarity of traditional construction contracts and of the New Engineering Contract. *International Journal of Project Management*, 15(4), 255–61.
6   CIB (1997) *Partnering in the Team*. Report by Working Group 12. Thomas Telford, Canada.
7   Bennett, J. & Jayes, S. (1995) *Trusting the Team: the Best Practice Guide to Partnering in Construction*. Centre for Strategic Studies in Construction, Reading Construction Forum.
8   Bennett, J. & Jayes, S. (1998) *The Seven Pillars of Partnering: A Guide to Second Generation Partnering*. Reading Construction Forum.
9   Pollington, T. (1998) *Promoting Good Clientship: The Work of the Construction Client's Forum*. GP Forum, No. 2.
10   Egan, Sir John (1998) *Rethinking Construction*. Report of the Construction Task Force to the Deputy Prime Minister, John Prescott, on the scope for improving the quality and efficiency of UK construction.
11   *Building* (1999) Industry chiefs doubt the impact of Latham. 12 February, 9.
12   *Building* (2001) Latham's legacy. 14 December, 41.
13   *Property Week* (2001) Tender-hearted. 30 November.
14   Latham, M. (2002) How far we've come. *Building*, 15 March, 33.

Chapter 12

Technology Foresight Report: *Progress through Partnership* (1995)

Stuart Green

12.1 Introduction

The Technology Foresight report *Progress through Partnership* was published in 1995 as a direct consequence of the White Paper *Realising our Potential* issued 2 years previously. *Realising our Potential* is widely acknowledged as a key landmark in recognising the importance of the UK research base and its contribution to national wealth. It succeeded in raising the profile of science and led directly to the formation of the Office of Science and Technology (OST) and the subsequent reorganisation of the research councils. The Science and Engineering Research Council (SERC) gave way to the Engineering and Physical Sciences Research Council (EPSRC). The key message of *Realising our Potential* was the need for a better alignment between Government, industry and academia.

The Technology Foresight process sought to emulate similar look-ahead methodologies established in the USA and Japan. The hope was to try to replicate the perceived success of 'Japan plc' with the development of a much stronger sense of mutual dependency among UK stakeholders. Technology Foresight unashamedly attempted to propagate the notion of 'UK plc'. Although the initiative was facilitated by Government, it was very much led by private industry. The Construction Panel of the Technology Foresight Programme was one of 15 sector panels that reported to a steering group. The 15 panels collectively consulted with over 10 000 people. The aim of the programme was to identify areas where new developments would yield benefits to the UK. The intention was that the results would then guide Government and industry decisions on funding research and development. The Construction Sector Panel focused on areas where it was thought effort was required to accelerate wealth creation and enhance quality of life.

The review offered in this chapter adopts an overtly critical perspective. The Technology Foresight report is initially placed within the context of the UK in 1995. The review then unfolds in accordance with the structure of the report. Finally, a view is offered on the impact of the report on the ongoing construction policy debate. Particular attention is given to the way in which the report has shaped the construction research agenda.

12.2   Context

12.2.1   *Political*

The political climate in 1995 was turbulent. The Conservative Party had been in power since 1979 and had resided over significant and lasting changes in the British economy. Under the leadership of Margaret Thatcher, Government had seen a significant and sustained shift to the political right. Policy sought to extend the domain of the free market throughout the economy in the cause of competition. Key dimensions included privatisation, deregulation, reduction of trade union power and lower direct taxes. The declared task of Government throughout the 1980s was to re-energise Britain by encouraging an 'enterprise culture'. Private sector management techniques were afforded an iconic status. The 1980s saw a significant decline in trade union power, exemplified by the defeat of the National Union of Mineworkers (NUM) in 1984. The age of globalisation had begun and the industrial conflicts that had characterised previous decades were seen to be a luxury that the UK could no longer afford. Government policy had shifted during the 1980s from adjudicating between competing domestic interests to trying to ensure that the UK was an attractive location for inward investment. Critics of Government policy claimed that the indigenous manufacturing base had been decimated through lack of investment. About 25% of manufacturing capacity had been lost during the recession of the early 1980s. Trade unions were excluded from influence and the infrastructure of cooperation between the state, labour and industry was progressively dismantled. This was well illustrated by the abolition of the National Economic Development Office (NEDO). The Government took no interest in promoting dialogue on economic policy between different social interest groups. Such coordination was anathema to the doctrine of the free market.

By the end of the 1980s, many commentators who had initially been sympathetic to the Government were arguing that the laissez-faire policies had gone too far. Mrs Thatcher appeared to be increasingly autocratic and dangerously out of touch with public opinion. The poll tax debacle was the final straw that caused the public to lose faith. In 1990 Margaret Thatcher finally lost the support of the Conservative parliamentary party and was replaced as Prime Minister by John Major.

Against all the odds, John Major won the general election in 1992. In many respects, his victory granted the doctrine of Thatcherism a new lease of life. The National Health Service (NHS) was reorganised to form an 'internal market'. Education was subject to endless change and the policy of privatisation gathered pace. The privatisation of the railways was initiated in 1993. The Major Government continued to be dogged by party divisions over Europe and accusations of Government sleaze. In June 1995 John Major resigned as party leader, thereby forcing a leadership vote. Although subsequently re-elected, the Conservative Party failed to unite behind him. Major's Government limped on to eventual electoral defeat in 1997.

12.2.2 *Economic*

The Technology Foresight report was published at a time when the UK economy was recovering from severe recession. Following the boom in development activity of the late-1980s, the output of the construction industry peaked in 1990 at around £55 billion. This fell to £46 billion in 1993 before recovering to almost £53 billion in 1995. The construction industry accounted for approximately 8% of gross domestic product (GDP) and 1 375 000 employees, 621 000 of whom were self-employed. Structural changes in the UK economy over the preceding decade had resulted in the decline of Government as an influential client. Nevertheless, the public sector accounted for approximately 30% of new construction in 1995 (excluding infrastructure). Private sector firms were becoming more involved in public infrastructure projects through the Government's Private Finance Initiative (PFI).

The UK economy of 1995 was characterised by low inflation and moderate, but steady, output growth. Although economic performance was slightly down on 1994, the overall diagnosis remained good. Unemployment accounted for 8% of the labour force, comparing favourably with European competitors. The previous 15 years of structural adjustment undoubtedly resulted in a more flexible economy. However, the Thatcherite experiment had failed to reverse Britain's economic long-term decline. It must also be recognised that Government economic policy throughout the Thatcher–Major years had been highly inconsistent. Despite the espoused monetarist rhetoric, damaging inflationary growth had been allowed during the mid-1980s. John Major's election victory in 1992 was rapidly followed by sterling's ignominious exit from the European Monetary System (EMS). Some £5 billion had been wasted in a vain effort to defend an unsustainable exchange rate. Ironically, the enforced 20% devaluation of sterling undoubtedly benefited the economy in the mid-1990s. Nevertheless, the Conservative Government's reputation for economic competence had been destroyed.

Whether or not the Thatcher years resulted in productivity improvements is still hotly debated among economists. Some emphasise the 'economic miracle' whereby the UK economy broke loose from restrictive practices and widespread overmanning. Others point to the possibility that extensive bankruptcies and closures enabled the survivors to utilise their capacity much more effectively, hence improving productivity[1]. History will ultimately judge the economic legacy of Thatcherism. Nevertheless, commentators such as Hirst[2] are clear that productivity was not helped by fresh investment in up-to-date processes. Investment fell sharply after 1979, increased from 1983 to 1990 and then fell drastically from 1990 through to 1995. In retrospect, the revenues from privatisation and North Sea Oil were squandered. A golden opportunity to invest in the modernisation of the UK's industrial base and its supporting infrastructure had been lost in preference to demand stimulation through tax cuts.

12.2.3   *Social*

While the economic legacy of the Thatcher years remains debatable, there is little doubt that they resulted in a much more divided society. Inequality returned to Victorian levels, with the development of a significant marginalised 'underclass'[3]. Whatever economic benefits were realised by the 'enterprise culture', they were not shared equally across society. The increase in UK wage inequality in the 1980s was matched only by the United States[4]. A significant burden of unemployment developed whereby 25 per cent of men of working age became 'economically inactive'[2]. There was also a marked growth in part-time and temporary employment, with a widespread reduction in employment protection. Homeless people became a common sight on the streets of Britain's major cities. The progressive weakening of family and community bonds resulted in unprecedented levels of crime. Relentless education reform damaged teacher morale and perpetuated bureaucracy. Academic achievement as an end in itself was downgraded in favour of the acquisition of knowledge and skills aimed at improving individual competitiveness in the market place. National institutions such as the Health Service appeared to be in terminal decline. The central state became progressively more intrusive, with a significant reduction in the autonomy of local government. Institutions such as the BBC and the universities became subject to much greater centralised control. The Thatcher–Major years also saw significant reductions in civil liberties, with restricted rights of public assembly and the abolition of the right to silence. Many of these trends continued under the subsequent Blair Government elected in 1997.

12.2.4   *Technological*

In 1995 the information and communications technology revolution was in full swing. Information technology (IT) utilisation in the construction industry increased significantly in the 1990s. However, implementation lagged behind more consolidated industrial sectors. The retail sector introduced electronic point-of-sale systems (EPOSs) for stock control, enabling 'just-in-time' ordering. The fragmented structure of the construction industry provided a significant barrier to such innovations. Computer aided design (CAD) systems were in common use, although very few construction firms had started to trade electronically. E-mail had been widely adopted within a short space of time, and electronic data interchange (EDI) and the World Wide Web were starting to have an impact. A few innovative construction companies were experimenting with technologies such as knowledge-based engineering (KBE), object modelling (OM) and virtual reality (VR). However, the impact of these practices on the construction industry as a whole was miniscule.

Despite the reduction in Britain's manufacturing capacity, there remained several areas of world-class technological innovation. Areas where Britain possessed a global competitive advantage included the production of specialised alloys, plastic processing, pharmaceuticals and biotechnology. Within the building materials sector, Pilkington's float process for manufacturing distortion-free flat glass was licensed throughout the world. Turnover in the computer sector in 1996 was £14 900 million[5]. While many computer firms were foreign owned, home-based firms such as Psion successfully concentrated on niche markets. British firms had also been involved in the development of a new generation of semiconductors. Within the communications sector, British firms were heavily involved in optical fibre communications systems. They competed successfully in high-technology sectors such as radio communications equipment, radar and thermal imaging equipment. The UK aerospace industry continued to support strategically important technological expertise. However, the aerospace sector experienced especially difficult times during the early 1990s. Reductions in defence orders and fierce international competition created doubts concerning its long-term prospects. While Britain's technological base and capacity for innovation were healthy in parts, there was little room for complacency.

12.3 Mission and methods

The overriding aim of Technology Foresight was to consider how the UK could best take advantage of opportunities to promote wealth creation and enhance the quality of life for UK citizens. In accordance with the 'trickle down' economics of Thatcherism, the two were assumed to be inextricably linked. The hope was that the panels would generate visions of the future that would lead to better-informed decision-making in both the public and private sectors.

The Technology Foresight Panel for the construction sector comprised 21 representatives drawn from business and academia (see Table 12.1). Big firms predominate, although representation was given to no less than six different universities. The major contractors were represented by John Mowlem, Laing Technology and Taywood Engineering. The latter two firms were later to face significant difficulties towards the end of the 1990s. The big suppliers were represented by Pilkington Glass, which has long enjoyed a quasi-monopolistic position in its relationship with the

Table 12.1 Membership of the Technology Foresight Panel on Construction.

Mr Herb Nahapiet	John Mowlem (Chair)
Prof. Patrick O'Sullivan	University College London (Vice-Chair)
Mr Charles Barber	Laing Technology
Prof. Peter Brandon	University of Salford
Dr Peter Bransby	CIRIA
Dr Roger Brown	Taywood Engineering
Mr David Button	Pilkington Glass
Dr Claire Carden	Building Research Establishment
Dr Linda Clarke	University of Westminster
Dr Keith Eaton	Steel Construction Institute
Prof. Malcolm Horner	University of Dundee
Mr Neil Milbank	Building Research Establishment
Mr Turlogh O'Brian	Ove Arup Partnership
Mr Jerome O'Hea	Colt Group
Ms Mary Rogers	Mary Rogers and Associates
Mr Peter Rogers	Stanhope Properties
Dr Robyn Thorogood	Department of the Environment
Prof. Barry Turner	Middlesex University Business School
Prof. Roger Wooton	City University
Prof. Barbara Young	University College London (Steering Group Assessor)
Dr Gladys Jones	Office of Science and Technology (Secretary)

construction industry. Given the fragmented structure of the construction industry, small firms were grossly under-represented. It is further noticeable that clients were few and far between. The minimum level of participation from public sector clients is especially noticeable in comparison with the Banwell and Emerson reports.

The methodology for the study involved a series of consultation workshops with members of an 'expert pool'. The workshops were attended by 203 people, predominantly from the private sector. The Panel devised a two-stage postal Delphi survey that was completed by a further 151 experts. Numerous trade associations and other organisations were consulted. There was a notable absence of trade union involvement. Furthermore, there was little representation from the small firms that collectively account for the majority of output.

12.4   The UK construction industry

The report provides a useful summary of the UK construction industry's strengths and weaknesses, albeit from a rather one-sided top-down perspective. Several long-standing arguments are repeated, with little in the way of fresh analysis. The UK is said to enjoy some of the lowest input costs in Europe in terms of labour while suffering some of the highest output prices. Reference is also made to the decline of national technological strength relative to the UK's competitors. It is interesting that the timeframe adopted for this comparison is 1967–91. This reflects a widespread assumption among management writers that the UK's decline in competitiveness is somehow due to the use of inappropriate management techniques. In contrast, political and economic historians[6–8] trace the UK's decline in industrial productivity to structural factors rooted in the nineteenth century. The report further refers to the low level of investment in R&D in the construction industry. This is contrasted with the relatively high level of research investment in the offshore oil and gas industry. The comparison makes no reference to the different structural characteristics of the two sectors. The offshore oil and gas industry is highly consolidated, dominated by large firms with significant barriers to entry. In contract, the construction industry is highly fragmented, dominated by small firms with low barriers to entry. Further comparisons are made with the pharmaceutical and automotive sectors, both of which have more in common with the structural characteristics of the offshore oil and gas industry than they do with construction. Given these differences, the usefulness of comparing respective levels of investment in R&D is highly questionable.

   The report criticises the strength of the construction research community on the basis of its relatively poor performance in the 1992 Research Assessment Exercise (RAE). Major weaknesses are seen to exist in management research and 'process-oriented' work generally. The report considers that this supposed weakness was in part beginning to be filled by managed research programmes such as the *Innovative Manufacturing Initiative* (IMI), the LINK *Integrated Design and Construction Programme* and through the development of the Department of the Environment (DOE)/industry *Whole Industry Research Strategy* (WIRS). The expression 'process-oriented' would seem to reflect the pervasive influence of business process re-engineering (BPR) in the management rhetoric of the mid-1990s. The early years of the IMI were influential in imposing BPR-type constructs onto the construction research agenda[9]. While it is possible to argue this did indeed fill a 'gap' in the research base, it is equally possible to argue that the narrow focus on process improvement served only to reinforce the industry's obsession with instrumental rationality. That the rhetoric of BPR was so readily accepted by the construction industry's 'experts' says much for its pre-existing allegiance to simplistic machine metaphors.

12.5   Challenges for the future

12.5.1   *A holistic approach*

The report emphasises that a successful construction industry is vital for the UK to create wealth and improve the quality of life. A successful industry is seen to be one that generates continuing profits over the long term. However, profitability alone is not considered enough. A 'holistic' approach is advocated, whereby the industry needs to be seen to be acting responsibly in terms of its effect on the environment, society and the workforce. There is no suggestion that any regulation or enforcement might be needed. This would of course be anathema to the 'enterprise culture'.

12.5.2   *Key challenges*

Six key challenges for the future were identified:

1 *To reduce costs, add value and sharpen international practice.* The mantra of cost reduction seems to be an ever-present element of construction industry policy. Exhortations for 'major improvements in productivity, cost and quality' are common. The ways in which such improvements are to be achieved are noticeably absent. What is apparent throughout the Technology Foresight report is the repeated rhetorical device of linking 'reduced costs' to 'added value'. Such a linkage serves to deny use of the word 'value' to those who feel uncomfortable with the relentless focus on cost reduction. In the same way, the language of quality has been appropriated by the same managerial discourse.
2 *To pay greater heed to environmental and social consequences.* As was the case with the previous exhortation to reduce cost, the report could hardly advocate paying *less* attention to environmental and social consequences. The assumption that the industry will respond to such public concerns of its own volition seems somewhat misplaced.
3 *To strengthen technological capability.* The report points towards the increased technological intensity of the industry in terms of automation of construction processes and the use of IT. It foresees a greater convergence of construction and manufacturing in the future to produce 'modular and prefabricated components and subassemblies'. The call for a greater use of standardisation and the need to modernise the technology of construction are highly consistent with the dictates of BPR.
4 *To improve education and training.* Little information is provided on *how* education and training are to be improved, other than a plea for versatile people who are able to draw from a wide range of technical specialisms.

The paradox of demanding versatility on the one hand and an uncritical acceptance of 're-engineering' on the other passes without comment.

5 *To upgrade existing buildings and infrastructure.* The report refers to the decay and obsolescence of the nation's infrastructure, highlighting the need to upgrade buildings, structure and infrastructure in an affordable manner. The construction industry is considered to have the key to the problem, a view that seemingly ignores a long-term failure to invest in public infrastructure.

6 *To re-engineer basic business processes.* The report advocates that the 're-engineering of basic business processes to provide 'lean', rapid and effective performance is now commonplace throughout other industries but rarely occurs in construction'. The introduction of such 'advanced business techniques' is exhorted as a matter of urgency. There is no recognition that the continued imposition of simplistic machine metaphors may contribute to the 'bad attitudes' and 'adversarial culture' that industry leaders repeatedly decry.

12.6 Opportunities and obstacles to progress

The Technology Foresight report concedes that the above six challenges cannot be properly addressed without considering the opportunities for, and obstacles to, progress. However, the arguments are once again less than convincing. The onus for change is laid at the feet of anybody other than industry leaders.

12.6.1 *Opportunities*

The listed opportunities repeat some of the key challenges. The first is to apply advances in information technology (IT) and communications more coherently and comprehensively. The second is to change 'education and training programmes so as to improve the effectiveness' of those entering and those already in the industry. The third is the development of newer or cheaper materials and processes. The use of the word 'cheaper' seemingly displays a rare rhetorical slip from the managerialist newspeak that dominates elsewhere. The concept of 'cheaper' is usually translated as 'value-added'. Also of interest is the way in which education and training are both linked to the notion of 'effectiveness'. There is no recognition of the role of the professional institutions and universities in protecting educational standards from the short-term demands of the marketplace. The report reflects the spirit of Thatcherism in undermining the notion of 'professionalism' in favour of the rigours of the marketplace. Repeated calls to make university education more 'relevant' have resulted in the increased dominance of narrow instrumental rationality in both undergraduate and postgraduate courses. Such trends sit ill at ease with the need to provide versatile graduates who are capable of innovative ideas that challenge accepted conformities.

12.6.2   *Obstacles*

The listed 'obstacles to progress' provide a timely reminder of the limitations of the free market that seem strangely out of place. The short-term fiscal policies of Governments and private sector organisations are both seen as problematic. This provides an obvious contradiction to the free-market ideology that prevails in other parts of the report. A further obstacle is presented by the fragmented industry structure, comprising a large number of small firms organised in temporary coalitions to address individual projects. Likewise, the reference to the industry's highly fragmented and casual workforce sits ill at ease with the Government's preceding 15 year assault on trade unionism in the cause of labour market 'flexibility'. The report is on more familiar ground when arguing that the procurement procedures and standard specifications used by large organisations and Government are a disincentive to innovation. Brief comment is made on the absence of feedback mechanisms in the industry before the usual criticism of British secondary and tertiary education. Education apparently increasingly accentuates greater specialisation rather than integration across disciplines. It is interesting to note that, when discussing 'obstacles to progress', no mention is made of training. Whereas 'education' provides an obstacle, 'training' provides only a challenge.

12.7   The engines of change

12.7.1   *Prioritisation*

Following the identification of 'opportunities and obstacles to progress', the Foresight consultation process was enacted with identified 'experts' to produce nine main priority areas, comprising four 'engines of change' and five 'key opportunities'. These collectively account for the report's main conclusions.

12.7.2   *Promoting learning and learning networks*

An organisation's only sustainable competitive advantage is said to lie in its ability to learn. The learning metaphor provides a welcome change from the machine metaphor that dominates elsewhere. The report states that education and training should cater for people's need to learn, to pass on knowledge and to adapt effectively to change. The recommendations include the establishment of a forum of 'leading educators and industrialists' to establish education policies and curricula to produce 'world-class constructors'. Existing mechanisms involving the professional institutions are clearly judged to be unsuitable. The Higher Education Funding Council is then exhorted to provide preferential funding to monitor and implement the forum's recommendations. Why preference should be given to this new forum over any others in existence is unclear. The EPSRC and Department of the Environment (DOE) are further encouraged to support the creation of 'learning networks' that embrace the different disciplines and technologies of construction.

12.7.3 *Reaping the benefits of the information revolution*

Emerging information and communications technology is held to promise 'immense gains in productivity and quality'. The report further suggests that IT can help improve supply chain integration. A case is made for the wider use of VR techniques during the early design stages. Other countries are apparently investing heavily in such technologies while the UK is lagging behind. Within the 10 years prior to the publication of the Technology Foresight report, the use of IT in the UK construction industry increased exponentially. The claim that IT can improve supply chain integration is undoubtedly true. It must also be recognised that the effective use of cross-organisational IT systems is dependent upon an established climate of trust. However, too many firms have experienced punitive supply chain management regimes that require them to deliver too much information for too little return. Furthermore, too many integrated supply chains means that small, occasional clients have little choice other than to deal with the industry's second division. The 'information revolution' alone will have little impact on the extensive fragmentation of the construction sector. While particular niches within the industry have invested heavily in IT, the barriers to widespread application cannot be overcome simply by raising awareness.

12.7.4 *Establishing a favourable fiscal regime*

The report identifies that UK fiscal policy is less favourable to investment than is the case in Japan and Germany. Unsurprisingly, the report is very short on advice on how the situation could be improved. It is widely recognised that endemic economic short-termism hinders total life costing and social and environmental profiling. While this is undoubtedly a market failure, the problem is deeply entrenched and by no means limited to the construction industry. The authors of the report seem to sense the futility of challenging the long-standing dominance of the City of London over the needs of British industry. The recommendations are that research councils should fund research into the relationship between changes in fiscal policy and its effects on the economy and construction investment. It is further recommended that the results should be disseminated widely. In the knowledge that nothing can be done to challenge the historical global orientation of British finance, it seems rather tepid to recommend further research.

12.7.5   *Creating a culture of innovation*

The authors of the report clearly feel more comfortable addressing the issue of innovation than they do with fiscal policy. The following sentence illustrates their enthusiasm:

> '[a] flourishing resource base needs to be built up in the UK, which is constantly refreshed and enriched by research and cross-fertilised by ideas from different disciplines and cultures.'

The report recommends funding for research into innovation and the establishment of Chairs of Innovation in university construction departments to encourage multidisciplinary research. Recommendations directed at industry bodies include the plea to promote a culture that is more favourable to innovation. Companies are further exhorted to appoint innovation directors, and clients are tasked with including companies' 'innovation profiles' as part of their supplier assessment profiles. In reality, relentless Government support of big business and the corresponding assault on trade unionism and the public sector demonstrated remarkably little respect for 'different disciplines and cultures'. The construction industry's regressive recipe of human resource management consistently falls short of encouraging a 'flourishing resource base'[10, 11]. Of course, the last thing that established industry leaders want is a workforce teeming with innovative ideas. It is much safer to pressure central government to encourage innovation through tax breaks and grants for investment in R&D. Perhaps the starkest assumption is that all parties benefit equally from the innovation process. Given increasing trends of wage inequality this would seem unlikely. The authors of the Technology Foresight report did not apparently share widespread public concern regarding 'fat cat' salaries in corporate boardrooms. In comparison, the incentives for innovation available to employees were very low indeed.

12.8   Key opportunities for increasing wealth creation and the quality of life

12.8.1   *Customised solutions from standard components*

The rationale for standardisation returns to the familiar themes of 'vital to reduce costs' and 'huge potential savings'. The proposal is for a 'set of reusable components and systems' that can be assembled in a variety of ways.

The inspiration is clearly provided by the automotive sector. Sensitivity to the dangers of standardisation is demonstrated by a list of technical requirements that includes aesthetic and cultural diversity, connectivity, low maintenance and environmental acceptability. The organisational and cultural requirements include the need for a 'fundamental, industry-wide reappraisal of design philosophy'. Who exactly was going to lead this reappraisal is not made clear. The need to learn from previous attempts at component standardisation is clearly acknowledged, seemingly in reference to the unhappy history of off-site fabrication in the 1960s.

The recommendations include a call for the DOE and EPSRC to establish a joint research programme with industry on the supply of customised solutions using standardised components. The DOE and industry are further exhorted to identify obstacles to standardisation in the form of regulation, always a safe target for criticism given the prevailing ethos of the time. Manufacturers are further encouraged to take a lead in developing key standard components, seemingly in the expectation that orders will be forthcoming. Once again there appears to be no envisaged role for the professions.

12.8.2   *Applying business processes*

The rationale for applying business processes repeats previous exhortations. While BPR is not mentioned explicitly, its influence is evidenced by the call for a 'step change' rather than previous incremental improvements. The proposed solution has the unpalatable flavour of a mission statement:

> '[a] vibrant and up-to-date modern construction industry providing a comprehensive 'best-in-class' (with low cost but high value) service to clients in the UK and throughout the world.'

The technical requirements include the much greater application of 'business processes' to the whole construction project delivery process. The storyline of BPR is further echoed in the requirement for a 'redefinition' of information needs and a 'redesign' of organisational relationships. Organisational and cultural requirements include the 'cost efficient' integration of suppliers within the process and the application of 'management and organisational techniques' for the purposes of achieving better cooperation and close integration. Such comments display a remarkable faith in the application of management techniques. It is further taken entirely for granted that people are compliant, predictable and willing to be programmed in accordance with the requirements of a rationally designed system. The possibility of employee intransigence as a result of the failure of previous top-down Taylorist management initiatives is not recognised.

Specific recommendations include the call that the EPSRC and Economic and Social Research Council (ESRC) should fund multidisciplinary research into the application of 'improved business processes'. The conclusions of this research would seem to be predetermined by the accompanying comment that 'business process analysis should be applied more vigorously to improve the efficiency and effectiveness of the construction industry'. There is a further call for a LINK programme to fund collaborative projects examining the application of techniques such as 'benchmarking' and 'total quality management'. The implication is that such management techniques are founded upon a scientifically respectable body of knowledge.

12.8.3   *Adopting a 'constructing for life' approach*

The rationale for 'constructing for life' hinges on the importance of operating costs as opposed to capital costs. The proposed solution is for building and civil engineering projects to maintain a proper balance between capital and operating expenditure while seeking to increase client and user satisfaction throughout the whole life of construction. The importance of life-cycle costing has long been recognised in the literature and academic curricula[12]. The problem is that neither contractors nor clients had previously taken such techniques seriously. One reason that clients ignore 'constructing for life' approaches relates to the perennial problem of fiscal short-termism. Contractors ignore them because they have little incentive to do otherwise. The technical requirements listed by the report seem to miss both of these key issues. Much is made of the importance of 'automated, non-destructive performance measurement devices' and 'powerful, user-friendly information repositories'. The report once again demonstrates a resilient faith in the ability of instrumental techniques to overcome deep-rooted structural barriers. The recommendations emphasise the need for industry clients to work with the research community to develop intelligent databases of measured performance and maintenance costs. There is a further call for prototypes of new forms of building. No recommendations are made regarding the difficulties of predicting future user requirements.

12.8.4   *Benefiting the environment and society*

Construction is said to have a wide-ranging and often unacceptable impact on the environment and society. Comments regarding the 'hidden social costs' of construction seem especially out of place given the tone of the report elsewhere. The proposed solution calls for a more holistic approach involving a 'comprehensive procedure to identify total environmental and social costs and to weight the benefits of alternative solutions to construction problems'. This would seem to be a somewhat utopian vision that ignores the previous widespread failure of cost–benefit analysis. Managers are seemingly required to take all possible future scenarios into account before making a decision. The technical requirements include a strange mix of cost/benefit accounting methodologies coupled with the use of smart sensors to monitor and improve health and safety. Further mention is made of 'durable yet recyclable materials' and the use of devices to provide 'real-time' performance displays of structures and external fabric. The recommendations include a call for a new research programme by the EPSRC/ ESRC/Natural Environment Research Council (NERC) to develop holistic techniques for social and environmental benefit analysis and 'whole life' assessment of the performance of constructed facilities. Yet again the proposed solution relies on the development of instrumental techniques. In contrast to other areas, there are no grandiose calls for the industry to 'change its culture'.

12.8.5   *Creating a competitive infrastructure*

Infrastructure is seen to be essential to the efficient functioning and economic progress of developed and developing countries. However, the emphasis strangely lies on developing countries and the associated opportunities for UK exports. Little attention is given to the long-standing lack of investment in domestic infrastructure. The Thatcher years saw widespread privatisation of public utilities in the cause of improving competitiveness. While some of the early privatisations were undoubtedly successful, critics claimed that national assets were being undersold. It is equally clear that Government throughout this period had little sympathy for public transportation systems. Infrastructure policy ultimately culminated in the privatisation of Railtrack, whose subsequent failure did much to dent the iconic status of private sector management techniques. None of this detracts from the obvious expertise of the UK engineers and their success in exporting their skills overseas. The orientation of the Technology Foresight report towards seeking profits from overseas markets in preference to domestic investment provides an interesting microcosm of the causes of the UK's economic decline[13]. At the time of writing, UK commuters would gladly sacrifice a 'competitive infrastructure' in favour of one that met the minimum requirements of safety and reliability. In this respect, the authors of the Technology Foresight report seem to have suffered from the same myopia that afflicted those in Government.

12.9   Impact

The Technology Foresight report was undoubtedly a child of its time. It represents a significant shift from previous reports in reflecting the elevated status of private sector ideology. Its rhetoric is a direct product of the 'enterprise culture'. The discourse is unashamedly optimistic and claims to speak on behalf of the construction industry as a whole, giving scant attention to the existence of conflicting interests. The differing needs of the construction sector are emasculated by an upbeat message of 'we're all in it together'. Conflicts between different interest groups are suppressed in the cause of UK plc. There is no room for diversity in the brave new world of Technology Foresight.

The overriding assumption of the report is that the construction industry has performed badly because of poor management techniques. The solution is therefore simple. The industry has to develop an innovative culture while applying modern management ideas more vigorously. The report abounds with the optimistic language of 'managerialism'. It reflects a wider project by a managerial elite to impose its language on the construction industry. Given the chosen mechanism of consultation among a designated 'pool of experts', the report was perhaps inevitably doomed to reinforce the existing prejudices of industry leaders. Its greatest impact arguably lay in the way it shaped the subsequent policy debate. The imposed requirement for a better alignment between Government, industry and academia relegated research to a subservient role. Research had become an instrument of policy. The subtext suggested that academics had previously become wayward and needed to be bought back into line. In this respect, the Technology Foresight report was to prove influential. The new climate was one where research had to be relevant to the needs of industry. Research as a generator of knowledge was no longer valued. History is placed in the dustbin as preference is given to a bundle of ill-defined management techniques. Knowledge is subjugated to the narrow cause of efficiency. Lean thinking has arrived.

In looking towards the future, the report identifies that 'clients are becoming more informed and more demanding, and client-led innovation could well be a dominant force for change in the future'. In this sense, the Technology Foresight report is prescient of the subsequent client-led Egan report (see Chapter 13). The 'cult of customer' became dominant in popular management discourse during the 1980s and resonates with the rhetoric of the 'enterprise culture'[14]. An ideological climate had been created where the legitimacy of large clients to speak on behalf of the public good was taken entirely for granted. In shaping the construction policy debate around 'an alignment of interests', the Technology Foresight report served only to reduce the sphere of legitimate discussion. Research had been tamed to serve the interests of big business in the cause of 'relevance'. Progressively, the construction policy debate was reduced to the narrow domain of 'efficiency'. The long-term effects of this on the construction industry remain unclear. Some hope is still to be found among the 'adversarial attitudes' of the industry's workforce.

12.10   References

1   Blackaby, D. H. & Hunt, L. C. (1992) An assessment of Britain's productivity record in the 1980s: Has there been a miracle? In: *Britain's Economic Miracle: Myth or Reality* (ed. N. M. Healey). Routledge, London.

2   Hirst, P. (1997) Miracle or mirage?: The Thatcher years 1979–1997. In: *From Blitz to Blair: A New History of Britain Since 1939* (ed. N. Tiratsoo), pp. 191–217. Weidenfied & Nicholson, London.

3   Commission on Social Justice (1993) *The Justice Gap*. Institute for Public Policy Research, London.
4   Machin, S. (1996) Wage inequality in the UK. *Oxford Review of Economic Policy*, 12(1), 47–64.
5   Central Office of Information (1996) *Britain 1997: An Official Handbook*. The Stationery Office, London.
6   English, R. & Kenney, M. (eds) (2000) *Rethinking British Decline*. Macmillan, Basingstoke.
7   Hobsbawm, E. (1999) *Industry and Empire*, 2nd edn. Penguin, London.
8   Pollard, S. (1989) *Britain's Prime and Britain's Decline: The British Economy 1870–1914*. Arnold, London.
9   Green, S. D. (1998) The technocratic totalitarianism of construction process improvement. *Engineering, Construction and Architectural Management*, 5(4), 376–86.
10   Coffey, M. & Langford, D. (1998) The propensity for employee participation by electrical and mechanical trades in the construction industry. *Construction Management and Economics*, 16, 543–52.
11   Druker, J., White, G., Hegewisch, A. & Mayne, L. (1996) Between hard and soft HRM: human resource management in the construction industry. *Construction Management and Economics*, 14, 405–16.
12   Flanagan, R. & Norman, G. (1983) *Life-cycle Costing for Construction*. Surveyors Publications, London.
13   Coates, D. & Hillard, J. (eds) (1995) *UK Economic Decline*. Prentice-Hall, London.
14   du Gay, P. & Salaman, G. (1992) The cult(ure) of the customer. *Journal of Management Studies*, 29(5), 615–33.

Chapter 13
Rethinking Construction: The Egan Report (1998)

Mike Murray

13.1 Post-Latham/pre-Egan UK construction (1994–8)

The 4 year gap between the Latham[1] and Egan[2] reports saw a general election in the United Kingdom and the installing of a Labour Government in Parliament. Recollection of previous Labour Governments would perhaps have suggested a 'pump priming' of the economy through investment of national infrastructure works. Such a mechanism for creating wealth and employment was of course supported by this new Government, albeit in the guise of the Private Finance Initiative (PFI). The Labour Government strongly supported this method (as it continues to do today) of financing public procurement, but it was also the case that both internal clients, such as the National health Service (NHS) for example, and contractors had a roller-coaster ride: the moral and ethical dimensions (in the level of patient care and transfer of employment rights) caused concern to the British Medical Association which adopted a national anti-PFI policy in 1998, and the high cost of tendering (up to £1 million) affected contractors. Nevertheless, the PFI and its predecessor, the Public Private Partnership (PPP), within which the PFI remit falls, provided a backdrop to public works UK construction throughout the 1990s.

A further significant aspect of the post-Latham construction industry has been, and continues to be, the promotion of a non-adversarial industry culture. Egan and his task force considered Sir Michael Latham's 1994 publication to be a 'landmark report'. In particular, this report was the catalyst for the 1996 Housing Grants, Construction and Regeneration Act (known as the Construction Act) in May 1998. As with the PFI, the Construction Act has been both praised and criticised. Indeed, a survey conducted by the Constructors Liaison Group (CLG) in 2000 revealed that many of Britain's top 100 main contractors 'are continuing to flout the Construction Act'[3].

Further changes to the industry culture between 1994 and 1998 demonstrate a deeper concern for environmental issues within the industry that would later result in the concept of 'sustainable construction'. Who can forget the images of road protestors such as 'Swampy' on our nightly news? In addition, the industry's poor record on health and safety continued to be a popular topic of interest. The negative image of the industry presented by such conditions manifested itself in recruitment problems at both the management and operative levels in the industry. Romans[4] blamed such stigmatisation on the industry itself when he argued that UK construction actively encourages its 'sexist, racist and homophobic image.' However, perhaps the third most significant aspect describing the post-Latham construction industry is that of the development of cooperative project relationships – partnering. The Construction Industry Board (CIB), established on the recommendation of Latham, published a report[5] in 1995, *Partnering in the Team*, to develop and illustrate Latham's views of partnering. Interest in this topic continued to grow and spawned further reports[6-8] and a plethora of industry conferences and seminars on this topic. Many clients moved from one-off project partnering relationships to continued strategic partnering which in itself has laid the foundations for the concept of supply chain management.

The 4 year gap between these reports was marked by a significant growth in client power in the industry. The Construction Round Table (client organisations such as British Airport Authorities (BAA), Tesco, Railtrack and the Highways Agency) published an *Agenda for Change* in 1997 which set out targets for improvement in the design of facilities, the trading environment and the delivery process. The Construction Clients' Forum (the Highways Agency, NHS Estates and the National Housing Federation, for example) published a *Pact with the Industry*. Both organisations are now defunct with the launch of the Confederation of Construction Clients (CCC) organisation in December 2000. The CCC brief is to represent all clients regardless of volume, value or frequency of construction activity. In spite of such client power, problems with 'cowboy builders' continued to provide the media with many a horror story of both legitimate and illegitimate builders. The ease of entry into this industry is seen to be a major factor, and Holt[9] comments on top-end suppliers in construction as 'two blokes with a rottweiler and a white van!' The Construction Task Force (CTF) believes that partnering will provide the barrier to entry.

13.2   1997: a construction task force

The CTF was established by the Deputy Prime Minister, John Prescott, and the new Minister for Construction, Nick Raynsford, in October 1997. The person chosen to lead this ten-man group was the former BAA Chairman, John Egan. Egan was considered to be a tough and demanding taskmaster who had previously revolutionised the production process at Jaguar Cars and was the driving force behind the changes in procurement practices taking place at BAA. Other influential clients invited to come on board included Tesco (supermarkets), Whitbread (restaurants and inns), Nissan (car production) and British Steel. The group excluded contractors and only one trade union (GMB) was included. To this day, the client-focused composition continues to attract criticism. A significant member of the group was Professor Daniel Jones whose interest in lean manufacturing would influence the report's recommendations. The CTF was to report back to the Government at short notice (the research was undertaken, written up and published all within a 10 month period), and one critic argued that 'no new significant things [knowledge]' would emanate from the report[10]. Although the 1998 report is now commonly referred to as the 'Egan' report (and Egan cited as the author), it should be noted that a 'backbench' team led by Simon Murray (BAA Team Services Chief Executive) was actually responsible for the report draft which was to be presented to the task force sitting with Deputy Prime Minister John Prescott.

The report has spawned a new language and narrative within the industry. 'Egan' itself is used as both a noun (person) and verb (to improve the process of construction, to remove waste, to increase efficiency, to rethink construction), with disciples of Egan referred to as 'Eganites'. Moreover, the concept of rationalising the supply chain, whereby contractors select preferred suppliers who grow in size by hoovering up those competitors who do not make the tender stage, has been described as 'Eganomics'. It is no wonder, then, that the report has been labelled by some as 'evangelical', with the body responsible for promoting the report recommendations, the Movement for Innovation (M4I), labelled as a cult. Green[11], a construction academic, has been vocal on Egan's reform policy and has argued that such a cult status is akin to the Spanish inquisition – contractors cannot speak out against the agenda in public for fear of losing work from clients who support such change. The mechanisms for achieving such change are largely based on manufacturing ideology and as such have introduced further new vocabulary (lean construction, concurrent construction, supply chain management, and key performance indicators) into the construction narrative. Keeping abreast with the current management 'fads and fashions' is naturally time consuming, and the weekly construction press can often be seen to provide a 'bluffer's guide' to Egan speak. Indeed, *Building*[12] refers to the 'gospel according to St John' and challenges the readers by putting their faith to test by completing a questionnaire, results being measured on an 'Eganometer'.

13.3   The *Rethinking Construction* report

The report is 40 pages in length and contains a foreword by Sir John Egan and an executive summary for those who wish to dip into the main findings. The main body of the report is divided into six chapters which progressively move from Chapter 1 ('A need to improve') through to Chapter 6 which proposes 'The way forward'. The other chapters set out the CTF ambitions for UK construction (Chapter 2) and move on to describe how the construction process can be improved by benchmarking other industries (Chapter 3). Chapter 4 ('Enabling improvement') examines the cultural changes that are considered necessary to enable such improvements, with Chapter 5 looking specifically at the improvements in efficiency and quality that can be made in the housebuilding sector.

13.3.1 A need to improve

The report considers that 'UK construction at its best is excellent', praises the industry's engineering ingenuity and design flair but stresses that it also needs to modernise. However, if we skip to the last page of the report, the language is a little less congratulatory and the magnitude of the challenge is clearly set out:

> 'We [the task force] wish to emphasise that we are not inviting UK construction to look at what it does already and do it better; we are asking the industry and Government to join with major clients to do it entirely differently. What we are proposing is a radical change in the way we build.'

Chapter 1 of the report argues that the industry 'recognises that it needs to modernise', although evidence to support this view may be considered anecdotal given the slow pace of change and innovation in this industry. One may also ask which sectors/stakeholders recognise the need to modernise. The clients who participated in the drafting of this report clearly think so, but what of the future of construction law practitioners? Particularly given that the task force talks of an 'end to [a] reliance on contracts' in Chapter 4 of the report. The problems that are deemed to require a 'make-over' are those that have beset the industry for decades and have all been previously identified in some manner by previous reports in this textbook: a low and unreliable rate of profitability, low investment in research and development, a crisis in training and clients who continue to equate price with cost and who select designers and constructors almost exclusively on the basis of a tendered price.

The CTF acknowledges that one of the most striking things about the industry is the number of companies that exist. The Department of the Environment, Transport and the Regions (DETR) has compiled a statistical register which shows some 163 000 companies listed, most apparently employing fewer than eight people. Such fragmentation is viewed as both positive (flexibility to deal with economic cycles) and negative because of the industry's reliance on subcontractors and the temporary nature of project teams, with no continuity of knowledge on a project-to-project basis. On a positive note, the report considers that both project and strategic partnering arrangements, as advocated by the Latham report, have influenced project performance, with capital costs and project programmes being significantly reduced, albeit for experienced clients such as Tesco (supermarkets) and Argent (commercial developer). Chapter 1 also refers to the adoption of techniques learned from other industries (benchmarking, value management, just-in-time and concurrent engineering) while suggesting that standardisation and preassembly would improve the industry's performance by speeding up construction time, lowering capital cost, reducing the need for skilled labour and encouraging a zero defect culture. Indeed, Egan suggests that manufacturers should make interchangeable components that fit together like a 'Meccano set'[13]. A performance improvement tool developed by the Building Research Establishment (BRE) and known as CALIBRE[14] is recommended as a means to eliminate waste and improve value-adding activities in the construction process. It is no surprise to learn that clients such as BAA, McDonalds and Tesco have all used CALIBRE on projects recently.

13.3.2   *Ambitions for UK Construction*

The CTF conducted a review of the changes that manufacturing companies (British Steel and Toyota, for example) and service companies (grocery outlets and offshore engineering oil and gas suppliers) have made over the years and concluded that they were only achievable owing to a series of fundamental changes, and are just as applicable to construction. Figure 13.1 shows that *committed leadership* is required to drive forward an agenda for improvement and that this will require cultural and operational changes throughout all parts of organisations that wish to change. The CTF does, however, appear despondent given that, after meetings with many managers in the UK construction industry:

> 'We have yet to see widespread evidence of the burning commitment to raise quality and efficiency which we believe is necessary.'

The second fundamental driver of change is that of *a focus on the customer* and involves removing activities in the manufacturing process that do not add value from the customer's viewpoint, which are classified as waste and eliminated. It is suggested that the construction industry tends not to think about the customer (either the client or the consumer) but more about the next employer in the contractual chain. The third driver of change is seen as resolving the detrimental aspects of construction's largely sequential process with its associated multistakeholder (and argued by the task force as manifesting itself in a lack of commitment to the long-term success of a project) arrangement. Thus, a need *to integrate the process and team around the product* is necessary. A *quality-driven agenda* forms another driver of change, and the report considers this not only to include zero defects but also delivering projects on time and to budget and innovating for the benefit of the client. The removal of waste, with its connotations of 'fit and lean' is also recommended to include not only waste in materials and construction on site but also design waste. Only if this is done can UK construction exceed customer expectations.

The last driver of change involves a *commitment to people* which is seen not only to mean decent site conditions, fair wages and care for the health and safety of the work force but also to include a commitment to training and development of these employees. The report concludes that much of the construction industry fails to realise that its people are its greatest asset. Indeed, Egan himself[15] stated some months after the publication of the report:

'I don't think you can be a businessman unless you understand behavioural things. You always have people as your basic raw material.'

Although it is understood that Egan is apparently 'well-meaning', the reference to 'raw material' does evoke a sense of people being merely a resource to be manipulated by management as it sees fit. It also raises interesting parallels with the personnel management field, where recent discussions have focused on the use of the term 'human capital'.

The report sets targets for improvement and proposes that the construction industry 'set itself clear measurable objectives'. This was deemed essential given that the CTF did not find any evidence to suggest that the industry had made the 30% improvement in productivity called for by Latham in 1994. The report sets out the scope for improvement in predictability, cost, time and quality, based on evidence derived from leading clients and construction companies in the UK and USA. The improvement figures set out are thought to be achievable year on year. These are: a 10% reduction in capital costs; a 10% reduction in construction time; a 20% increase in predictability (number of projects completed on time and budget); a 20% reduction in the number of defects on handover; a 20% reduction in the number of reported accidents; a 10% increase in productivity (value added per head); and finally a 10% increase in turnover and profits of construction firms. The CTF comments that these are minimum rates of improvement and that it expects the best UK construction companies to surpass them.

13.3.3 *Improving the project process*

The task force argued that construction can 'seek improvement through re-engineering construction and by learning as much as possible from others who have done it elsewhere'. Moreover, they conclude that construction is no different from manufacturing and that, in common with this industry, the overall construction process can be subdivided into four complementary and interlocked elements. These are product development, project implementation, partnering the supply chain and production of components (see Fig. 13.1). This integrated project process is seen as a radical change from the traditional model of project delivery and is intended to link together the complete supply chain. The concept of lean thinking is also introduced in this chapter, and the task force believes that its adoption in the UK construction industry will lead to sustained performance improvement. Lean philosophy is of course based on Womack & Jones's textbook *Lean Thinking*[16] which in itself is based on earlier work by these authors and published in 1990[17].

13.3.4   *Enabling improvement*

Chapter 4 of the report emphasises that a substantial cultural change will be required if UK construction is to meet the improvements that the report asks for. The welfare facilities available to the workforce on construction sites come under scrutiny and are described as 'typically appalling'. Furthermore, the report places an emphasis on improving construction's health and safety and notes that it is the second worst of any UK industry. The CTF also considers that training and quality are inextricably linked and questions whether this industry has the right skills to improve productivity. In addition to promoting good-quality training with the possibility of multiskilling being encouraged, the report asks for a culture of continous (lifelong) learning to be adopted in the industry. Egan and his colleagues also recommended that every trained construction worker should hold an identification card that would list the employee's skills, and that clients should only employ contractors who can demonstrate this. Although clients such as BAA already demand this of contractors, the industry continues to use 'black market' labour. In particular, significant numbers of asylum seekers/immigrants can be found working on London sites. At the time of writing, the new Construction Minister is holding talks with the Home Office over plans to legitimise illegal (skilled) immigrants so that they can apply for work on construction sites[18].

This chapter also covers such topics as buildability and argues that the separation of design and construction has led to both construction problems on site and buildings that do not perform satisfactorily when in operation. Suppliers and subcontractors should be involved at the design stage and whole life costs should be an important consideration for such integrated teams. Clients must also play their part by accepting that projects require time for adequate preconstruction planning. This section of the chapter also alludes to the management concepts of knowledge management (KM) and organisational learning (OL) when it asks that the experience of completed projects be fed into the next one.

Reference is also made to the benefits (greater efficiency and quality) of adopting standardised components and processes within the industry, and it is argued that this need not lead to poor aesthetics or monotonous buildings. The use of technology as a tool to support the cultural and process improvements is also recommended, and the benefits that can be gained by adopting information technology to aid the design process are outlined. The elimination of waste through standardisation and careful use of technological capability is, however, seen to be secondary to making changes to the industry's culture.

The use of long-term relationships or alliances should be adopted widely, and these relationships should be based on trust and not rest on contracts. Most importantly, the report suggests that, where relationships between constructor and employer are based on mutual interdependence, then 'formal contractual documents should gradually become obsolete'. The adoption of strategic partnering would also, it is argued, reduce the need for tendering and focus clients on requesting value for money rather than lowest tenders. Such conditions are however seen to require a rigorous qualification framework where suppliers are set quantitative performance targets but with the added advantage (assuming both constructors and employers can change from an adversarial culture) of open-book accounting.

13.3.5   *Improving housebuilding*

The Egan report looks specifically at one sector of the construction industry. Housebuilding is said to be affected by several significant factors that distinguish it from contracting or civil engineering projects. Examples are given as the planning and building control regulations which impinge on the level and location of the activity, the 'one-off' disaggregated client base in private housebuilding and social policy changes that result in uncertainties in the housing association/local authority market. The report recommends the establishment of a housing forum that would assist cross-fertilisation of innovation between the public and private sectors. The main improvements are, however, thought to be better achieved through the social housing sector because a few major clients commission these projects.

13.3.6   *The way forward*

Chapter 6 calls for commitment from major clients, from the construction industry and from the Government. The major clients on the CTF agreed to lead the way by supplying demonstration projects that would be used to develop and illustrate the ideas in the report. In addition, the CTF sow the seeds for the Movement for Innovation (M4I) when they propose the establishment of a movement for change. It is, however, regognised that construction is a 'deeply conservative industry', and great will from all sectors will be necessary if a change of style, culture and process is to lead to more than just a series of mechanistic activities. The establishment of a knowledge centre/hub to disseminate best practice is recommended, and the DETR's Best Practice Programme is recommended for this role.

Drivers for change	Improving the project process		Annual target for improvement	
Committed leadership			Capital cost	−10%
Focus on the customer	Product development	Partnering the supply chain	Construction time	−10%
			Predictability	+20%
Product team integration			Defects	−20%
			Accidents	−20%
Quality-driven agenda	Project implementation	Production of components	Productivity	+10%
Commitment to people			Turnover and profit	+10%

Fig. 13.1   The Egan agenda for change.

13.4   Delivering change

In total, seven Government bodies were established to deliver the change demanded by the report (see Fig. 13.2). The Housing Forum is responsible for promoting innovation in housebuilding; the Local Government Task Force is seeking to improve efficiency in local government and the Government Construction Clients Panel is seeking to improve central government procurement policy. The other four bodies are the Client Project Managers who will concentrate on improving clients' performance, Respect for People, which seeks to improve the working conditions in the industry, the Movement for Innovation (M4I), which is identifying best practice and promoting change, and the Construction Best Practice Programme (CBPP), the remit of which is to raise awareness of the need for change within the industry.

13.4.1   *Movement for Innovation (M4I)*

Egan and his team clearly understood that the CTF should not become a 'talking shop' (critics would appear to levy this charge at the majority of pre-Latham reports) and that some mechanism was required to deliver change within the industry. Indeed, as one industry practitioner comments, 'a lot of this has been said before, but this time its official and in big lights'[19]. The M4I was established in November 1998 with a mission to lead radical improvement in construction through value for money, profitability, reliability and respect for people. This is to be achieved through dissemination of best practice and innovation. The core feature of the M4I was to become the

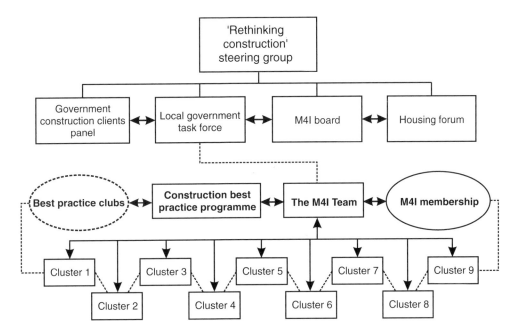

Fig. 13.2 *Rethinking Construction*: a framework for delivering change.

'demonstration project' programme. Industry has been encouraged to offer projects for inspection that show a clear culture of innovation. This has indeed been a contentious issue so far, with many project organisations wishing to promote and market themselves through this mechanism. Projects that are accepted by the M4I as demonstration projects require project teams to measure their performance using the KPIs and report back to the M4I (with guaranteed anonymity) via a secure Internet link, as well as attend regional cluster meetings. This initiative has, however, had teething problems, with many projects showing no clear innovation (other than supplier innovation) and late or non-return of KPI data and/or non-attendance at regional cluster meetings. Egan himself is also reported to be frustrated with the slow speed at which best practice techniques and principles (collected from demonstration projects) are being disseminated[15]. This is perhaps not a criticism of the work undertaken by the M4I and the CBPP but of an industry culture that uses information as not only a currency of power but also a competitive advantage. Moreover, it was revealed[20] that, on several of such demonstration projects, site staff at foreman level and below were completely unaware of the project's status. This is of course contradictory to Egan's philosophy of workforce empowerment which seeks to give all blue-collar and white-collar employees a stake in projects.

13.4.2 *Eganisation of construction*

A fundamental principle that emanates from the report concerns the application of knowledge derived from other industrial sectors to construction. The CTF believes that construction can learn best practice from manufacturing and the service industry and challenges construction to re-engineer its frameworks. In particular, the sequential nature of the traditional construction process is questioned and it is recommended that the skills and knowledge of suppliers and constructors (taken to mean subcontractors) can be captured during the design and planning stage of projects by the creation of an integrated project process. The report explicitly defines the working culture that it expects to emanate from such an arrangement:

> 'The key premise behind the integrated project process is that teams of designers, constructors and suppliers work together through a series of projects, continuously developing the product and the supply chain, eliminating waste in the delivery process, innovating and learning from experience.'

To date, the most successful example of such an integrated process is that of the Ministry of Defence (Defence Estates) *Building Down Barriers* initiative. The two pilot projects (swimming pools and gymnasiums) involved contractors Laing and Amec working as 'prime contractors'. The Tavistock Institute were advisors to this initiative, which involved charting the success of the supply chain integration. This was based on a 'clustering' framework whereby the projects were divided up into discrete parts (substructure, swimming pool, etc.) and all those organisations that were required for the supply of materials, design and construction of each part could work together[21].

13.5 Impact of Egan's report

According to Sir John Egan[22], 'Normally what happens to Government reports is people flick through them and throw them in the bin'. Alan Crane, M4I Chairman, has stated that:

> 'We have had so many reports over the past 30 to 40 years. All of them have had some effect but in the end have gained dust on the shelves.'

Clearly, the Egan report has not suffered such a fate, and it is perhaps the most widely cited reference in the construction industry today.

Given that a central tenet of Egan's philosophy is to 'measure and improve', one can be sure that evidence of improvements in the industry, or at least change, will not be difficult to uncover. However, we must not

assume that any (or all) impacts have indeed been desirable for each stake-
holder in the construction process. The rationalisation and consolidation (*vis-
à-vis* mergers and acquisitions) that have come to dominate the post-Egan
years will not have been favourable for all. Such consolidation within both
the contracting and housebuilding sectors is perhaps necessary as a defence
against overseas competition. This predicament may not, however, be unto-
ward for clients who possess the Egan bible.

Improvements made to public procurement practices have recently come
under scrutiny from the National Audit Office (NAO). *Modernising
Construction*[23] reviews the impact that the Egan agenda has had on four
public organisations: NHS Estates, Defence Estates, the Highways Agency
and the Environment Agency. It is concluded that, owing to only a small
amount of information being available, quantifying any benefits is difficult,
but all organisations predict significant savings in construction costs and
improvements in quality. Egan has also criticised Government departments
for not doing enough: 'some tried hard but got nowhere'[13]. Moreover, the
report notes that the CBPP estimates that it has reached only 9% of the
working population in the industry through its programme of seminars and
workshops. This can hardly be considered a 'groundswell', and cynics may
indeed say that it is largely preaching to the converted. The Auditor
General's report also comments:

'It is yet difficult to determine whether the improvements in process and
growing awareness of best practice have started to deliver improvements
in value for money as recommended by [the Egan] report across the spec-
trum of industry and clients.'

One particular sector of public spending, local authorities, has been largely
left behind by the Egan reforms, although more forward-looking councils
have embraced many of the recommendations and have registered demonstration projects with the M4I. The Local Government Task Force has,
however, published a toolkit on implementing Egan best practice, and this
has been posted to all 386 local authorities in England.

Top industry figures have also been critical of the reform taking place in
the industry. Sir Martin Laing commented that 'the recommendations of the
Egan and Latham reports are beginning to take root but at a very slow pace'[24].
Linda Clarke, a respected academic from the University of Westminster, has
compared social housing projects in the UK with other European countries[25]
and concludes that

'though the UK construction industry demonstrates high openness and
readiness to change and innovate, the question raised is whether it is able
to transform itself by example.'

Such a comment clearly questions the long-term benefits of the M4I demonstration projects. Constructions' supply side is also under scrutiny. A report commissioned by the CBPP, *How Small and Medium Size Construction Companies Measure Up*[26], found that many suppliers to the construction industry are inefficient in their deliveries and supply far to many defective products. Despite such findings, client BAA is planning to prefabricate up to 70% (therefore a greater reliance on suppliers) of the Heathrow Terminal 5 project off-site. Given that BAA's framework suppliers are working for a prestigious client, one might assume that they would embrace such a move. However, BAA's supply chain manager notes that all suppliers are not so Egan compliant: 'we have had a mixed reaction to it all'[27].

13.5.1   *KPIs: a definitive measure of industry improvement?*

The Egan report acknowledges that the pre-Egan industry had no performance indicators and that solid data on a company and project performance in terms of efficiency and quality are 'hard to come by'. Egan therefore proposed an industry-wide performance system, which would enable clients to differentiate between the 'best and rest', providing a rational basis for selection, and to reward excellence. Thus, the birth of key performance indicators took place, with a launch in May 1999. The ten KPIs were developed from the CIB's 1998 framework in association with the DETR, M4I, CBPP and the CCC (Eganisation has also spawned many acronyms!) and are akin to a league table concept whereby each organisation is able to compare itself with industry (such a knowledge bank does of course rely on prompt feedback from those organisations partaking in the scheme) and thereby identifying its strength and weaknesses. Seven are applied on a project-by-project basis: construction cost, construction time, cost predictability, time predictability, defects, product satisfaction and service satisfaction. Three indicators look at company performance: profitability, productivity and safety. The Egan challenge is therefore for sustained improvement in efficiency and quality rather than a one-off cosmetic approach.

Although all projects listed with the M4I as 'demonstration projects' are required to submit their KPIs, *Construction News*[28] reported that a significant number failed to do so. Anecdotal evidence suggests that many project teams see this listing as somewhat prestigious but fail to devote the time required to data collection and analysis. Cynics may, however, argue that, when a project fails to live up to expectations (i.e. quality, cost and programme problems), then one would not wish to advertise this, even when offered guaranteed anonymity. The paradox here is of course that these results should be used to inspire/motivate or simply 'perk up' project teams to do better next time. The paranoia that tends to exist in the industry, particularly at contracting level (argued here to be a result of many years of competitive tendering) does, however, not encourage openness. Indeed, the former chief executive of the

CIB, Don Ward, requested that clients should not use contractors' and consultants' KPI scores as a basis for selection. Rather, they should ask 'how are you using your KPIs to improve'[29].

The publication of KPIs for the year 2000 (67 out of 170 demonstration projects registered with the M4I) has revealed that these projects are safer, (this being despite HSE statistics showing that, for the 12 months to 31 March 2001, there were 114 fatalities in the construction industry) have fewer defects and are providing better client satisfaction than the industry norm. In addition, these projects are said to be passing the year-on-year target improvement figures set in the Egan report in all but three of the headline KPIs: cost, profitability and predictability of design cost. The Housing Forum has 100 registered projects, and results from 18 of these again show that they are outperforming the housing industry norm in all but two KPIs. The use of prefabrication and modular building techniques with a need for a steep learning curve have apparently resulted in both profitability and time predictability being lower than the industry average. These results have prompted the M4I Chairman to declare[30]

'the doubters in the industry have now got very clear evidence of the business case for adopting the principles of *Rethinking Construction.*'

13.6 Discussion

Egan has been awarded a 'second bite at the cherry' with his position of leadership in the Strategic Forum, established in July 2001. The CIB was disbanded owing to clients withdrawing support in the belief that this organisation was failing to promote their strategic business goals. One may question Egan's motives in accepting such a position, given that in an interview with *Building* in May 2000[31] he responded to the question 'Do you still care about *Rethinking Construction*? with the reply 'apart from a sort of observing interest, no'. Egan has nonetheless painted a rather gloomy forecast for UK construction if his mission fails by suggesting that this industry may be completely foreign owned in the future. On an optimistic side, the alternative would see the development of 'some absolutely superstar, world-class competitors growing up in the UK'. This is the mission of the Strategic Forum, to build on the work of the CIB and incorporate the demands of *Rethinking Construction*. As Egan notes, 'it's work in progress'[15]. However, in a draft foreword to the follow-up (*Accelerating Change*) to the 1998 report, Egan voices his concerns[32]:

'I have been greatly impressed by the response to *Rethinking Construction*. But I am frustrated that the rate of take-up has not been as rapid as it should have been.'

In spite of such frustration, the report has changed the industry. No other report reviewed in this book has spawned as much interest or had so much written about it. This is perhaps due to the Government's commitment in its sponsorship through the DETR, and the appointment of a Minister for Construction. The commitment by the Government (construction's biggest client, withan annual spend of £45 billion per annum) to construction has, however, been put under question with the responsibility for construction moving to the Department of Trade and Industry (DTI) and the replacement of a dedicated Minister for Construction with a Minister for Energy and Industry. Latham was indeed vocal on the loss of the Construction Minister role, describing it as a 'great shame'[33] and a 'complete dog's breakfast'[34].

The challenge that the construction industry now faces is perhaps better defined by the previous report *Constructing the Team*. One might ask which team, given that construction's concept of a project team is somewhat skewed by a familiar set of team attributes, which often manifest themselves in a negative manner. The multidisciplinary and multiorganisational nature of project teams is of course the most obvious. The progress made in keeping team participants together on a project-to-project basis (many clients, contractors and design teams now request 'named individuals' to work with) may also be problematic, as anecdotal evidence suggests a high turnover of staff throughout the industry. Perhaps more importantly, it would appear that the only report reviewed in this textbook that acknowledges the blue-collar workforce as stakeholders in the team is indeed *Rethinking Construction*. However, there is often a chasm between the respect demonstrated towards operatives (blue-collar workers) and that shown to white-collar workers, yet project delivery is equally dependent on good performance from everyone[35].

The first section of this chapter reviewed the developments in the industry between 1994 and 1998 and noted the inefficiencies in the industry. It is perhaps worth drawing this chapter to a close by briefly commenting on a project that demonstrates the innovative qualities of UK construction. The innovative £75 million 'Eden' project in Cornwall is the largest greenhouse construction in the world. The Sir Robert McAlpine–Alfred McAlpine joint venture project involved the construction of four geodesic domed greenhouse structures with a footprint equivalent to 29 football pitches and has been procured under a new engineering contract (NEC) with a guaranteed maximum price (GMP). The expected 500 000 visitors per annum will learn about man's relationship to plants, with the internal environmental conditions requiring Ove Arup to complete one of its largest computer modelling exercises. However, perhaps the sting in the tail for Egan (Adam bites into the apple so to speak) is that the design, fabrication and erection of the proprietary space frame system was all undertaken by a specialist German contractor. The second world-breaking feature of the project (the largest structure of birdcage scaffolding ever erected) is, however, the responsibility of a UK innovation, with Scaffolding Great Britain (SGB) employed in this role[31, 36]. The connotations of this German–UK comparison (skill, knowledge and ability) would not be lost on Egan given his reference to a UK construction market dominated by overseas firms!

13.6.1 The ideology of Egan

Egan's ideology seems appropriate for the twenty-first century world. Keeping it lean is all-important for businesses that wish to be competitive. Moreover, the influence that the city stockbrokers have had over the strategic direction of UK construction appears to be increasing. Egan's desire to generate a 'superstar' or group of superstar players in the UK is somewhat akin to the adage 'big business is very wise, let's get into enterprise' (Johnny Rotten). Indeed, Egan argues that the M4I demonstration projects are showing how to make money, how to satisfy customers and how to stop killing people[13]. Like Egan, Rotten and his colleagues encouraged radical reform, in this case within the popular music industry. However, just as 'old punks' grow old, so will the lifespan of Egan's report. Its momentum relies on innovation: innovation through cultural change, through product and process developments, innovation in projects and in strategic business goals, innovation through continuous improvement and innovation in adding value to clients, of course. To keep the ball rolling, the Strategic Forum is to publish what has been touted as a 'sequel' to *Rethinking Construction*. The report has been given the working title *Accelerating Change* and its publication is expected in the summer of 2002[37].

13.6.2 A final word!

It is fitting to end this chapter of the text, and indeed the concluding chapter of this review of Government reports, with a quote from Egan himself. It does perhaps emphasise the need to remember the 'KISS' principle, that is, 'Keep it Simple Stupid'. Egan, in discussing the need to eliminate waste from the construction process, appears to remove the science from the process of construction: 'It's about the elimination of waste, it's not Einstein'[38]. More recently, in referring to the actual report, he said[13]:

'It's all pretty straightforward. You don't need to be a brain surgeon to get into the improvement process.'

13.7   References

1   Latham, Sir Michael (1994) *Constructing the Team*. Final report of the Government/industry review of procurement and contractual arrangements in the UK construction industry. HMSO, London.

2   Egan, Sir John (1998) *Rethinking Construction*. Report of the Construction Task Force to the Deputy Prime Minister, John Prescott, on the scope for improving the quality and efficiency of UK construction.

3   *Contract Journal* (2000) Contractors still defy law. 19 July, 3.

4   Romans, M. (1995) CIOB president claims industry 'revels' in sexist and racist image. *This Week's News*, 23 June.

5   CIB (1997) *Partnering in the Team*. A report by Working Group 12. Construction Industry Board, Thomas Telford.

6   Bennett, J. & Jayes, S. (1995) *Trusting the Team: the Best Practice Guide to Partnering in Construction*. Centre for Strategic Studies in Construction, Reading Construction Forum, UK.

7   Bennett, J. & Jayes, S. (1998) *The Seven Pillars of Partnering: A Guide to Second Generation Partnering*. Reading Construction Forum, UK.

8   Barlow, J., Cohen, M., Ashok, J. & Simpson, Y. (1997) *Partnering, Revealing the Realities in the Construction Industry*. The Policy Press, University of Bristol.

9   Holt, R. (2000) Imagine there's no site labour. *Construction News*, 7 January, 22–3.

10 Gray, C. (1998) Task force faces flak from expert. *Contract Journal*, 28 January.

11 Green, S. D. (2000) Industry 'brainwashed' by Egan's message. *Contract Journal*, 19 April, 1.

12 *Building* (1999) Are you a believer? 17 December, 36–7.

13 Contract Journal (2001) Eganomical thinking. 3 October, 16–17.

14 Vassos, C. (2001) CALIBRE – The toolkit for facilitating world class performance in the UK construction industry. CIB World Building Congress, April, Wellington, New Zealand, paper HPT 02.

15 Egan, Sir John (2001) Troubleshooter. *Construction Manager*, September, 17.

16 Womack, J. P. & Jones, D. T. (1996) *Lean Thinking*. Simon and Schuster, New York.

17 Womack, J. P., Jones, D. T. & Roos, D. (1990) *The Machine that Changed the World*. Rawson Associates, New York.

18 *Building* (2001) Wilson in talks to legitimise illegal immigrants. 2 November, 11.

19 *Building* (1998) Is Egan on target? 24 July, 18–23.

20 *Contract Journal* (2000) Firms ignore workers' input on M4I projects. 16 February, 1.

21 Nicolini, D., Holti, R. & Smalley, M. (2001) Integrating project activities: the theory and practice of managing the supply chain through clusters. *Construction Management and Economics*, 19, 37–47.

22 *Building* (2001) Egan on Egan, 10 August, 18–19.

23 NAO (2001) *Modernising Construction*. Report by the Comptroller and Auditor General. National Audit Office, London.

24 *Contract Journal* (2001) Industry giants sound warning. 18 July, 4.

25 *Building* (2001) Wrong to pin hope on Egan, says report. 8 June, 14.

26 CBPP (2001) *How Small and Medium Size Construction Companies Measure Up.* Construction Best Practice Programmme, UK.

27 *Construction News* (2001) Jumbo prefab for Terminal 5. 8 November, 1.

28 *Construction News* (2000) M4I warns firms over KPI figures. May, 2.
29 *Contract Journal* (2000) Don't use KPIs as score cards' says Industry Board. 12 April, 9.
30 *Contract Journal* (2001) Demonstration projects outperform the industry. 11 July, 1.
31 *Building* (2000) Gardeners world. 24 March, 43–9.
32 *Building* (2002) Industry must do better, says new Egan report. 5 April, 12.
33 *Contract Journal* (2001) Blair's reshuffle leaves construction voiceless. 13 June, 1.
34 *Building* (2001) Minister calms industry after Whitehall snub. 15 June, 11.
35 Respect for People Working Group (2000) *A Commitment to People 'Our Biggest Asset'*. DETR, London.
36 *New Civil Engineer* (2000) The greenhouse effect. 18 May, 22–4.
37 *Construction News* (2001) Blue Circle boss to update Egan report. 20 December, 2.
38 *Contract Journal* (1997) The force is with us. 29 October, 14–15.

Chapter 14
Conclusion

Mike Murray & David Langford

14.1   Introduction

The journey in the previous chapters has been long. Over half a century exists between the 1944 Simon report and Egan's 1998 *Rethinking Construction* report. Such a journey can be charted by an examination of the construction industry through this period. From Lloyd George's 'homes for heroes'[1] to Blair's 'cool Britannia', the construction industry has played a significant role, be it economical, political, social or indeed environmental, in the nation's livelihood. The industry has been exposed to much public and media attention during this period, and to some extent 'overexposed' owing to its site-based nature. Unlike the manufacturing sector, construction has not been able to hide under the factory roof, away from the gaze of the media and public, and sheltered from the vagaries of the British climate. As such, many of the problems that have been identified in the preceding chapters have been attributed to construction's uniqueness. Undoubtedly, the temporary multiorganisational nature of project work, with guarded demarcation in occupational roles (both professional and non-professional), has contributed to construction's operating culture.

14.2   The long view 1944–2000

The reports reviewed in this book are considered to be those that have had most influence on the industry, its people and organisations. However, what is clearly evident from such a historical review is the significant number of recurring themes that span the 54 years between the first and last report examined. Critics may argue that this disappointing scenario has been self-inflicted and that the root cause lies with construction's conservative attitude to changing its lack of innovation and a reluctance to challenge the status quo. Indeed, Flanagan *et al.*[2] in reviewing several of the reports discussed in this textbook (Simon, Emmerson, Banwell, Tavistock, Latham), note that the opportunities presented by the reports offer a 'salutary message'. However, they recognise that, in spite of the good intentions, very little radical or fundamental change has taken place in either the product or process during the past 50 years. In the generality of the long view this may be the case, but important and specific changes in production methods and procurement practices can be detected. For example, the Wood report sought to gain efficiencies in public sector housing procurement by a preference for serial contracts so as to maximise production runs for industrialised building units. This mechanism for protopartnering between public sector client and proprietary building firms has been developed some 25 years on by public agencies such as the Defence Estates and the National Health Service which seek partnering arrangements that may be considered as a derivation of serial contracting.

While the use of prefabricated units did not last at the time, and serial contracting failed to become popular (certainly the Paulson affair, in which a Local Government architect was jailed for corrupt placement of building contracts, would have stifled the development of public contracts let by serial methods), the theme of standardisation reappears time and time again. Indeed, the Egan report commends McDonald's burger chain for its ability to erect a 'restaurant' within 24 hours. Given the concerns about the McDonaldisation of society the jury must surely still be out on the effect that such standardisation will have on the quality of the urban landscape.

Comparing the manufacturing and the construction industries in terms of their policies towards standardisation, it may be seen that one, manufacturing, is moving away from standardisation and economies of scale and to methods driven by flexibility and economies of scope using limited equipment to produce a wider range of products. At the same time the construction industry is still being urged to adopt a Fordist approach to production. This is hardly a future orientation.

After procurement and standardisation, a third theme is the integration of the professions. In the long view it may be seen that the changes in procurement methods and professional payment systems have shifted the patterns of relationships. In respect of procurement, highly fragmented models dominated until the mid-1970s. The process was sequential and lengthy. As procurement became more integrated, so hard professional boundaries softened and integrated professional practices became more common. Nowadays, integrated professional teams are formed for projects which can be made up of several independent practices who regularly work together or seamless integrated firms.

The liberalisation of trade in the professions was also a powerful driver of change. Until the mid-1970s, the professions prospered by a combination of fixed-scale fees and cost escalation on construction projects. When the Monopolies and Mergers Commission denounced scale fees as anticompetitive, the professions had increasingly to fee bid their work and so were exposed to the vagaries of the market. Such pressures stimulated professional integration.

It may be expected that the professions will become even more integrated. Since the 1990s there has been increasing concern on the part of employers and the educational establishment that fragmented education leads to a fragmented set of professions. Darbyshire[3] noted that the supply of professional skills required did not match the changing needs of the construction industry almost 10 years later. The Fairclough report[4] added to the momentum for change.

Such reforms, along with educational and professional integration of European qualifications in the construction sector, could shape the future of the professions. Already the European Council for Building Professionals has developed a pan-European postgraduate qualification in Construction Management. It will be equivalent to a Master's degree and will be launched in 2002. The mutual recognition of qualifications and free movement of the professions around Europe and beyond will presage further integration.

The long view suggests that this integration will proceed apace and that the benefits will be felt in a less disputatious industry that is concerned with shared problem solving and mutual respect for the contribution made by each party. The challenge is for the industry to engage in culture change to allow such developments to take place. They do, however, suggest that the Latham report has started to reverse this trend. Perhaps the tale by Rudyard Kipling *A Truthful Song* is only representative, then, of a pre-Latham and Egan construction industry[5]:

> I tell a tale which is strictly true,
> Just by way of convincing you
> How very little, since things were made
> Things have altered in the building trade.

One might find agreement with the sentiment above in spite of obvious signs that this industry has been progressive over the past 50 years. Clear progress has been made in areas such as the mechanisation of the building process, the use of information technology (IT), the use of new materials and creative and imaginative architectural designs, to name a few. However, it is perhaps the social and cultural structure of the industry that has been slow to embrace change. Apportioning blame to any section of the industry for such faults may not be so easy. Government intervention in this industry has been piecemeal, with no apparent long-term strategy. The use of construction as an economic barometer between the 1950s and 1980s is well understood, and successive Governments have tinkered with public construction works investment as a means to boost the UK economy. The resultant stop–go policies have done much to deter those seeking secure employment in this industry, with graduates and school-leavers alike seeking employment elsewhere. This is perhaps the most obvious sign of an industry without a coordinated mission.

14.2.1   Cultural changes in the workforce

The working culture of the industry can be seen to have changed. The strong protestant work ethic that prevailed in the late 1940s and 1950s during the 'rebuilding Britain' period was to develop into an era of militant strike action during the 1960s and 1970s and this in turn resulted in massive job losses during the recessionary period of the 1980s. The development of the self-employed worker (with 714/SC60 tax certificates) during the 1980s reflected the Thatcherite principle of 'let the market decide', and to some extent today's twenty-first century cowboy builders have their origins in this era. Such obvious change may not be so apparent in the white-collar construction worker. Professional institutions such as the Royal Institute of British Architects (RIBA), the Royal Institution of Chartered Surveyors (RICS), the Chartered Institute of Building (CIOB) and the Institution of Civil Engineers (ICE) have steadfastly protected the interests of construction's professional elite and to a large extent their own interests over that of the paying client. Hindle[6] has argued that the industry suffers from a 'log jam' of professions and that such 'professional congestion' has a detrimental impact on the industry. It is argued here that little has changed in the manner in which these professions have worked together over the past 50 years, in spite of the obvious change in procurement systems (traditional, design and build, management) and contracts. As such the divide between such blue-collar and white-collar workers is much akin to the recurring theme that is explicitly noted in many of the reports – the separation of design and construction.

14.2.2   Technological developments

The technological aspects of the industry have also changed, and some have come 'full circle'. The prefabrication techniques adopted in the 1940s and 1950s, used to house those returning from World War II were also adopted in the high-rise buildings during the 1960s. The Ronan Point disaster in 1968 did, however, raise questions about industrial building techniques, and Granada Television's 1980s exposé on timber frame house construction (Barratt) led to a lack of confidence in prefabricated house building. However, confidence in prefabrication returned, and, in the 1990s, clients such as BAA and McDonald's were championing its use. Indeed, in 2000, the Peabody Trust (housing association) commissioned the UK's first multistorey housing project to be entirely factory built.

The adoption of advanced technology is of course crucial if construction is to demonstrate its commitment to the DETR's 2000 report *A Strategy for More Sustainable Construction*[7]. The call for a construction industry that is socially and environmentally responsible only became an important issue in the 1990s, which reflects the UK Government's commitment to international 'green' legislation. The sustainability issue is useful in contrasting the reports examined in this text. The industry has moved from consumption in the 1950s to conservation in the 1990s and the current UK Government has encouraged this move by imposing tax payments on construction's use of landfill sites and fresh aggregate.

14.2.3   *Clients and customers*

The overwhelming need to repair bomb damage and rebuild Britain in the 1940s and early 1950s meant that the public client offered the industry the greatest opportunities during this era. Powell[8] notes that the Government had great capacity to influence decisions to build during this era, and that until the early 1950s private promotion was displaced from its former pre-eminence by state intervention. Indeed, Hillebrandt[9] notes that Government intervention in the industry was most evident by its control of building permits that gave priority to essential work. These were discontinued in 1954, allowing private developers to have some influence on the industry. The slum clearances and rapid erection of high-rise accommodation by local authorities in the late 1960s and 1970s combined with the investments being made by central government in infrastructure, also meant that the public client continued to provide significant work. Powell[8] does, however, confirm the decline in state intervention when he notes that, between the mid-1970s and the late 1980s, the value of public sector work plummeted from around 60% to approximately 25%.

By the 1990s the regular build clients were becoming vocal about the service they were receiving from the construction industry. Two influential client bodies were established, the Construction Round Table (CRT) and the Construction Clients Forum (CCF), and these bodies were seen to demand an improved service from the industry. Central and local government was also to re-emerge as an important client. The growing number of Private Finance Initiative (PFI) hospital, school and prison projects offered contractors a steady income stream over a concession period, the marked difference being that those contractors were offering a service to the client rather than a product. Other clients such as the Ministry of Defence (MOD), British Airport Authorities (BAA) and supermarkets such as Asda and Tesco started to offer serial contracts to preferred contractors. This method of procurement has seen a marked rise in the number of negotiated contracts, something that would no doubt please Simon, Emmerson and Banwell, in spite of its protracted nature.

14.2.4   *Procurement systems*

The history of the reports examined in this book can be charted by the role of the key design and construction personnel involved in projects and the procurement systems used by clients to commission projects. Between the

1940s and 1970s, the architect was considered by most clients to be the key professional expert, and the use of a traditional procurement system was the recognised method of commissioning a building project. During the 1980s, the use of design and build procurement started to increase, and the concept of a novated design team became popular with some clients. This naturally resulted in a change of power between design and construction personnel. The adoption of management contracting and construction management was popular in the 1980s and 1990s, and this was to result in the creation of contractors who did not build! The growing number of specialist contractors employed as work package contractors reflected this trend, and companies such as Sir Robert McAlpine, Bovis and Mace were employed by clients to manage and coordinate projects. The almost universal use of project managers in the industry today also reflects the changing attitudes of clients who are convinced, or coerced into believing, that they are essential. Indeed, the exact role of this relatively new profession in construction is often questioned. Alsop[10], a chairman of an architectural practice, was quoted as saying: 'I don't see the point in their existence. Twenty years ago they weren't around and we [architects] took on the responsibility'.

14.2.5   *Recurring themes*

As previously noted in the introductory chapter, the reports examined in this text have a number of recurring themes that reflect an industry inflicted with long-term illness. The content of many of the reports are strikingly similar and, indeed, the contributors in Chapter 5 (Banwell report 1964) commented to us that the number of similarities with the 1998 Egan report were striking. What is evident, however, is the change in language that spans the reports. The concepts of supply chain management and lean construction are all too evident in the forerunners to Egan, but without the appropriate buzzwords. This is an important aspect of an industry change. As a means to combat an apparent volatile and unpredictable market, construction has become more reliant on advice delivered from management consultants and gurus. To some extent this relies on creating a new industry paradigm, one where management is dominant and the use of a new language indicates a commitment to radical change. Sims[11] has argued that the most famous buzzword of all, partnering, has been hijacked by consultants and corrupted by contractors. Furthermore, many of the new ideas are repackaged common sense. This new language may indeed be the building blocks for the twenty-first century construction industry, but critics would argue that too few within the industry can 'walk the talk' and even fewer can 'talk the talk'.

What of the recurring themes? Tables 14.1 to 14.3 have been used to provide a quick reference so that one can contrast and compare the reports. It is not exhaustive but is intended to reflect the nature of each report. The tables are based on that of Rogan[12] who used a typology with 29 descriptors

Table 14.1 Recurring themes 1944–98: procurement (after Rogan[10]).

Report	Procurement			
	Contractor selection	Nomination	Serial tenders	Partnering
Simon 1944	Selection should be based on character and ability, responsibility and pride in work; in return, fair remuneration for good service should be given	Indefinite relationships between the general contractor and various subcontractors nominated by the architect. Where work is an integral part of the design, STCs must be placed in advance of main contracts	London County Council's sliding fee scale should be used for continuous programmes of work	Negotiated contracts with the chosen builder have the advantage of establishing a relationship based on confidence, assuring consultation with the architect and builder. The disadvantage is that this may be more expensive
Phillips 1949	—	Only in exceptional cases of highly specialised work should the architect nominate a subcontractor or obtain separate tenders for work	—	—
Emmerson 1962	The system of placing building contracts should be reviewed comprehensively; open tenders unacceptable	Nomination is needed in appropriate circumstances	Serial contracts should be used as they reflect the need for collaboration between designer and subcontractor	Efficiency in building operation is dependent on the quality of the relationship between building owner, professions, architect, surveyor, engineer and contractor and subcontractor
Banwell 1964	Repeats Simon, requirements are 'character and ability, responsibility and pride in work, with fair remuneration for good service'. Open tendering should be removed. Early selection need not preclude competition.	If early nomination is part of the specialist work, then the main contractor should also join the team early	—	Negotiated contracts should not be excluded in the public field; methods of contracting should be examined for the value of the solutions they offer to problems rather than orthodoxy

Table 14.1 Cont.

Report	Procurement			
	Contractor selection	Nomination	Serial tenders	Partnering
Tavistock 1965 and 1966	—	If the main contractor is nominated early in the building process, he can be party to the nomination of subcontractors (1965)	—	—
Large Industrial Sites 1970	Management contracting mode of procurement preferred, with contracts set up on a reimbursable and negotiated basis. Rigorous prequalification procedures should be in place	Clients would be better served by greater integration of manufacture and install arrangements for specialist equipment. This would certainly be by nomination	—	Encouragement for clients and contractors to 'partner' with the trade unions for mutual benefit of reduced stoppages and controlled casual labour
Wood 1975	Current practices, open competition 16%, select competition 65%, negotiation 14%, two-stage tendering 3%, serial 1%; percentage of completed contracts surveyed within 5% of contract sum, open 56%, select 58%, negotiation 66%, two-stage 82%; open tendering to be abolished	—	Serial tenders give feedback to the design team from earlier contracts; serial or continuity tenders should be used for house building and schools programmes that allow close collaboration. The disadvantage is that the contractor may not act as he did on his first contracts	Pure negotiation is appropriate in certain circumstances, but clients may pay more and it will take greater effort by the client to get value for money
NEDO 1983	Successful fast contracts were where the contractor was chosen not on price but on previous performance, with a willingness to accept the customer's urgent deadline	Temptation to nominate STCs for design and supply to reduce the workload on the designer may lead to disruption of the programme; incompatibilities of STCs identified too late, information cannot be incorporated in design	—	—

Cont.

Table 14.1 Cont.

Report	Procurement			Partnering
	Contractor selection	Nomination	Serial tenders	
NEDO 1988	Choice of the main contractor usually based on competition	The majority of contractors appointed the specialists 'named' or 'suggested' in the tender documents. The short time available to prepare for the site operations made it impracticable to look for alternatives	Many regular and major customers had established procurement paths, and the expectation of repeat orders motivated the industry	Where customers have established a firm and well-defined context for coordinating the contributions and responsibilities of all the main participants, including their own, project objectives can be accomplished in a spirit of confidence and partnership
Latham 1994	Should be a register of consultants and contractors for selection; tender list arrangements need rationalising; tenders should be assessed on quality as well as price	Many specialists would like to see a return to nomination, but only 11% of specialists' work is now nominated	—	Partnering with teamwork/win–win approach helped bring Sizewell NP on time and budget; partners must be sought through a competitive tender process
Technology Foresight 1995	—	—	—	—
Egan 1998	The industry rightly complains about the difficulty in providing quality when clients select designers and constructors on the basis of lowest cost and not overall value for money, clients do not benefit from having a new team on each project	—	—	Tools to tackle fragmentation such as partnering increasing, now used by the best firms in place of traditional contract-based procurement and project management; partnering saves 30% on cost and 50% is possible, as well as 80% in time. The use of integrated teams working on a series of continuous projects is important for learning to take place

Table 14.2 Recurring themes 1944–98: relationships (after Rogan[10]).

Report	Relationships			
	Client	Main contractors	Subcontractors (STCs) and suppliers	Teamwork, trust and cooperation
Simon 1944	Owners should insist on the completion of design; owners are led to believe the only way of knowing the cost is through tenders; the responsibility for inefficiencies lies squarely with the clients – they can prevent it if they wish	Not to invite or receive tenders that are the result of open advertisement, owning to the greatly varying standard of performance of the work, also not to invite more than a suitably listed number of tenders	Nomination in advance, with STCs helping to prepare the design; not to issue designs prepared by specialists for tender to other specialists; STCs should be allowed to offer alternatives	Efficiency and success are dependent on an honest desire for cooperation
Phillips 1949	Government clients are different from the ordinary building owner and this places on them a unique responsibility; these departments should shoulder their responsibility and play their part. The building owner must pay more for the building if variations are introduced	—	Representatives of subcontractors submit that nominated subcontractors should have more protection against loss owing to insolvency of the principal contractor	It is of vital importance that all concerned in building operations should be animated by the right spirit and should take the right attitude towards their responsibilities in relation to the industry; they must cooperate fully and wholeheartedly in everything that helps to improve efficiency and cut out waste. Architects have the most marked influence on the efficiency of the industry
Emmerson 1962	More consideration needs to be given to the interests of the building owner; clients rely on the advice given by the architects as they often lack experience	—	—	A lack of confidence between the architect and builder, amounting at its worst to distrust and mutual recrimination, and at best to aloofness, causing inefficiencies which owners pay for
Banwell 1964	Clients who spend money on construction seldom spend enough time at the outset making up their minds	Endorsed by all in 20 years, not to invite or receive tenders that are the result of open advertisement, greatly varying standard of performance of the the work	STCs and nominated should be integrated to form a cohesive team; if STCs are appointed by the main contractor, the STCs should have all the facts	Relationships based on mutual confidence lead to seedy economical outcomes; the relationship to design and build must be improved by common education

Cont.

Table 14.2 Cont.

Report	Relationships			
	Client	Main contractors	Subcontractors (STCs) and suppliers	Teamwork, trust and cooperation
Tavistock 1965 and 1966	—	The main contractor, in undertaking the sponsorship of the construction team, has the function of optimising the mix of resources and skills called for; the main contractor needs to maintain a series of communications through the construction process (1965)	STCs contributing to the design will need to communicate instructions to the MC in his role as designer, but be under the control of the MC in his functions as an STC. It is often reported that this position causes communication problems during the construction phase (1965)	The setting up of the construction team is mainly the responsibility of the main contractor; factors such as personal acquaintance, goodwill or favours owed often come into play in the setting up of the construction team. There are wide discrepancies in social perceptions an opinions about the value of contributions to the building process among members of the building team (1965). To a degree, the industry has a tacit loaylty to present conditions for reasons of immediate partisan advantage (1966)
Large Industrial Sites 1970	Delays and costs overruns tend in part to stem from decisions taken by the client before work on site begins; critical choices for the client include the form of contract, the number of work packages to let, input in design, payment of damages and level client supervision; approval for client variations only where absolutely necessary	—	The greater the number of STCs, the more need there is for sophisticated management to coordinate the complex set of interactions on site (also, see nominations)	The poor industrial relations climate in engineering construction prompted the report, and it notes the conflictual nature of the relations between management and unions. The report set out to change this climate by engendering trust and cooperation through forging a new national agreement between the contractors and unions. This was to be underpinned by training in industrial relations for managers and union representatives

Table 14.2 Cont.

Report	Relationships			
	Client	Main contractors	Subcontractors (STCs) and suppliers	Teamwork, trust and cooperation
Wood 1975	Conspicuous failure of clients in this important role; clients provide a signal interface between client, designer and contractor; care in the selection of the design team is down to the client	Case studies provide evidence of successful projects designed by contractors	A large number of STCs who can contribute to design – greater use should be made of them; major STCs can be crucial to the success of projects	A construction project requires the collaboration of the parties; teamwork and thoroughness on the part of those carrying out the projects is equally important to organisation and systems
NEDO 1983	Most projects that went well were for experienced, well-inputted clients; customers must be organised to answer design queries authoritatively and at short notice; customers should be told of the consequence of changing their mind as work proceeds, with the results of cost and time penalties	Early inclusion before design being finalised may help programming and reduce problems; care should be taken in the choice of contractor; project arrangements that allow early, precise and integrated procurement and preparation for construction are needed	Properly managed STCs can lead to speedy and efficient construction; many delays are caused by poor coordination; STCs had the most far-reaching effect on construction times, only 10% of cases with total subcontracting had fast construction times, compared with 33% where direct labour was employed	Evidence was found that architects were failing to establish an effective dialogue with customers enquiring into expectations and standards. Contracts successfully complete in a short time are where good relationships, common objectives and mutual trust are established between the customer and STCs, particularly on fair payments
NEDO 1988	Customer has a key influence on the outcome of building projects; where the client was closely involved with a project the quality was generally good; industry's service to regular and major customers was better than that to occasional customers	Contractors rarely made assessments of the consequence of inadequate information, they accepted information as it stood; management resources often appeared inadequate, causing the site agent merely to respond to crises during execution of the works	M&E contractors criticised for poor organisation of design and rushed installation of complicated services	Architects should be taught about the building process, including compulsory periods on site and management training

Cont.

Table 14.2 Cont.

Report	Relationships			
	Client	Main contractors	Subcontractors (STCs) and suppliers	Teamwork, trust and cooperation
Latham 1994	Clients are the core of the process and their needs must be met; Government should commit itself to best practice clients	—	Joint code of practice for STCs should be drawn up, based on fair tendering and teamwork; STCs complain about onerous contract conditions from large engineering specialists to whom they are subcontracted	MC relations with STCs were very poor, with Dutch autioning common; most STC requirements were incompatible with the MC terms, yet the MCs acknowledged that the performance of STCs was crucial to the success of their own organisation. Promote a multidisciplinary education programme with cross-disciplinary experience
Technology Foresight 1995	Construction firms must achieve significantly higher levels of user satisfaction, in terms of linking the performance of a facility, building or structure to the needs of the users and investors, over the long run	—	—	The industry should set up mechanisms to ensure all players in the construction process are kept well informed and their activities fully coordinated by means of advanced information and communication technology (ICT)
Egan 1998	Clients do not always get what they ask for, value for money, free from defects, delivered on time; clients' immediate priority is to re-duce capital costs and running costs and improve quality; clients must also accept responsibilities for defective design, too often the clients are impatient; the public sector is the largest client	Cut-throat price competition and inadequate profitability benefit no one. The task-force wants a culture of radical and sustained improvement in performance; escape from the debilitating cycle of competitive tendering, conflict, low margins and dissatisfied clients	Need for a fully integrated supply chain critical to driving innovation and sustaining improvements; suppliers and STCs have to be fully involved in the design team	Form relationships in the supply chain, based on trust, and keep teams together; Designers need a more practical training and should wok in close collabora-tion with other participants in the design process. Partnering rela-tionships require interdepen-dence but will bring about long-term satisfactory arrangements

Table 14.3 Recurring themes 1944–98: performance (after Rogan[10]).

Report	Performance			
	Design and design sign-off	Prefabrication and standardisation	Quality	Value for money
Simon 1944	Other professionals now help in design; rushed design results in mistakes. Dates should be given on the programme for handing over detailed drawings	—	Industry is renowned for poor quality with wide price fluctuations	—
Phillips 1949	—	Considerable development in the prefabrication of buildings or part of buildings; prefabrication shows promise for the interior of the building rather than the shell. A development of standardisation that is receiving attention is that known as modular coordination	No evidence to show that bonused work is generally of inadequate quality	—
Emmerson 1962	There is inadequate preparation of plans before going out to tender; the process could be improved by greater use of standard drawings, bills of quantities, form of contract; in no other industry is the design so far removed from production; Clients are unaware of the advantages of having completed scheme designs	Substantial economies in school buildings through CLASP prefabricated systems with a continuous work programme, still allowing for variation in design. More standardisation is needed, to lower the expense of maintaining large stocks, to ensure sound quality and design and to reduce costs generally	The quality of work is dependent on the standards and disciplines set by the architect; open tendering is prejudicial to a firm that maintains high standards	—
Banwell 1964	Those who continue to regard design and construction as two separate phases are mistaken; in no other industry is the design so far removed from production (repeated from Emmerson)	Economies of scale through production line off-site assemblies where possible, with dimensional coordination and standardisation, this will not impair quality of performance	Clients should know and expect standards and quality at the outset; work, however small, should be carried out to a recognised standard	Insufficient time spent on the importance of value and time

Cont.

Table 14.3 Cont.

Report	Performance			
	Design and design sign-off	Prefabrication and standardisation	Quality	Value for money
Tavistock 1965 and 1966	The complaint of lack of design information at the contract stage is a hardy perennial in the building process; this raises the question of what is adequate documentation at this phase of the building process (1965)	—	The experience of the team has been of and industry in which misunderstandings, delays, stoppages and abortive work result in failures in communications, and the impressions of confusion, error and conflict (1966)	Obtaining value for money is constrained by the level of uncertainty in the building process; nothing contributes more to the industry's inefficiencies than uncertainty, since it provides the ideal environment for conflict (1966)
Large Industrial Sites 1970	Responsibility for design and design changes was a virtual issue for the committee. The fault line in the discussions was that many believed that the dependence of a functional specialist to make design changes cost the project in terms of delays. This view prevailed and project management were to have responsibility over technical issues	—	Causes of delay in large projects include late delivery of materials and plant, late design changes, low labour productivity and labour disputes, delays by subcontractors, faulty materials and workmanship. Trade training is erratic	The prompt for the report was the cost and time overruns in major engineering construction projects. These vicissitudes led to clients being distressed by the lack of 'value for money' in the UK engineering construction. Fees of inward investment dried up as a result of project overruns
Wood 1975	On large projects, coordination of the design team should be a separate function; whether or not a project gives value for money depends largely on the design team	The application of building systems will give the opportunity for superiority in methods and materials	Private clients are more aware of quality and added value, and therefore can often make variations to enhance value	—

Table 14.3 Cont.

Report	Performance		
	Design and design sign-off	Prefabrication and standardisation	Quality	Value for money
NEDO 1983	Work on site needs to be based on furnishing complete information, communicated clearly and in good time; difficulty in coordination the timing and details of design caused by STCs engaged through competitive tendering who were unwilling to invest time and effort into potentially abortive work when they had not been given an order. Complete freezing of design before construction starts is rarely feasible	The fast completed buildings were designed for simplicity and buildability, with particular attention paid to speed of construction	Speed of construction need not affect quality	—
NEDO 1988	Poor production drawings caused low quality finish, poor cost control and failure to meet completion dates; good sites were where designers put effort into planning; problems usually due to poor design or inadequate information. Late and incomplete information was a major frustration on many sites, with drawings arriving in dribs and drabs as work progressed; STCs contributed substantially to the design of buildings and were expected to guarantee their design	Specifications based on national standards produce better quality results. Specification is an integral part of design, yet it is of varying quality and comprehension and often technically irrelevant	Serious quality problems; mainly due to design information, through inaccuracies, late provision or failure to allow for construction needs; STCs' supervision and quality control of labour-only operatives often inadequate	—

Cont.

Table 14.3 Cont.

Report	Performance			
	Design and design sign-off	Prefabrication and standardisation	Quality	Value for money
Latham 1994	Effective management in the design process is crucial for the success of the project; design coordination before construction starts is impossible; the design team must offer the client a vision of the project that he can understand	Stanhope, a developer, is looking to use prefabrication and modular techniques to improve efficiency on site; McDonald's reduced cost and time in the UK by 60% in past 5 years using modular techniques	Clients have the right to expect high quality on projects; defects or failure in design cost industry £1000 million per annum; if clients are serious about quality, then they should only use QA assured firms	Identify methods of establishing what is best value for money and how to implement them
Technology Foresight 1995	—	Customised solutions from standard components should be adopted; huge potential savings in cost and time and an increase in quality from the use of standardised flexible components, manufactured off-site	—	The UK construction industry will be profitable through producing world-class products and service for markets at home and abroad, through making major and successive productivity improvements and through fostering and benefiting from an innovative culture, stimulated jointly by Government and industry
Egan 1998	Clients believe significant value improvement with cost savings is achievable where the majority of design work is complete before the construction phase starts; designers often fail to tell clients about product improvement, which is a problem for smaller clients; the use of dedicated design teams working exclusively on one design from beginning to end gives innovation in design and assembly; more prominence in the design and planning stage before anything happens on site; time spent in reconnaissance is not wasted; design needs to encompass whole life costs	McDonald's construct a restaurant in 24 hours; a survey of European housing sites reveals the use of a high degree of prefabrication and assembly, with efficiency put down to preplanning with suppliers and component manufacturers to minimise time spent on site. Standardised components and preassembly will improve quality and performance	30% of construction is rework; training and quality are inextricably linked; quality will not improve and costs will not decrease until the workforce is educated in skills and the culture of teamwork; quality is fundamental to the design process	Clients need better value from projects; value management can reduce costs by 10%; removal of non-value-added activities from the supply chain

to assess 14 Government reports. This typology has been condensed and the tables are split into three distinct areas covering the overarching themes of procurement (contractor selection, nomination, serial tenders, partnering), relationships (clients, main contractors, subcontractors and suppliers and teamwork, trust and cooperation) and performance (design, prefabrication and standardisation, quality and value for money). An absence of text in the tables indicates that the report concerned made no significant comment on the given subject. This is most evident under the 'Value for money' heading in Table 14.3, where it can be seen that several reports have not referred to this concept. Undoubtedly, the report authors were seeking to provide the industry (clients) with value for money, but it is recognised that the terminology is once again recent. The move from referring to 'value' rather than 'cheaper' is most evident in the 1990s when the use of value engineering and value management techniques became common in construction.

14.3   Final words

Construction is but one UK industry where Government intervention has an influence over its destiny. One mechanism that can be used to coerce and direct an industry is the publication of formal reports. This text has reviewed the most influential reports published by the Government on the construction industry over the past half-century. It has shown that successive reports continued to uncover the same industry ills time and time again. From Simon (1944) to Egan (1998), the industry has been characterised by fragmentation and the separation of design from construction. In latter years, the burgeoning number of specialist subcontractors should have smoothed out this barrier (i.e. constructor's input to design), but alas it has only complicated matters by requiring more extensive coordination and management during both the design and construction phases. Indeed, reading through this text provides much evidence of the black side of construction. Recommendations for improvements made in successive reports have been ignored or partially implemented and the industry has reformed at a pedestrian rate. How far has this reform to go to get the industry that clients crave? Internal, industry-level reformation has been prompted by each report. However, such reforms, in respect of greater integration of the construction process, greater investment in IT, research and development and staff development, have to be congruent with a wider environment in which the construction industry has to work. This external environment will be the enabler or brake in the process of change. Much will relate to the performance of the economy and how construction benefits or suffers. The fortune of firms in the industry is hostage to indicators such as interest rates, unemployment, inflation and economic growth. All of these factors have been encouraging for the industry in the early 2000s but questions remain over whether these conditions are sustainable in the long term. Interest rates are set to rise through the middle of the 2000s, and unemployment in the face of the global economic slowdown is set to increase as corporate performance wanes after its heyday through the 1990s. The prospect of continuing and extended war in the Middle East threatens oil price stability and therefore global economic performance. All does not augur well for the domestic construction industry. In spite of these external factors, current forecasts of growth through to 2005 suggest 2% increases in workload (at constant prices).

Against this background one has to ask what factors will shape the momentum that the most recent reporters, Latham and Egan, have started. Considering a few of the key issues that have dominated the change agenda over the last few years, we can conclude by evaluating how matters may play out.

14.3.1 Procurement

Developments in this area may be expected to shift the procurement debate to one of how supply chains are managed. This has implications for the structure of the industry. Larger firms will be expected to grow significantly in size, with a concentration on support services for clients, or to be investment companies with interest in infrastructure or property. This will squeeze the medium-sized players, and the largest firms will seek to integrate specialist contractors into construction projects. The mega-firm will of course be international in scope, where economies of scale can work for them. This will enable supply chain issues to be driven by sharing of expertise between the mega-firm and the 'boutique' specialist which are tied together by common design facilities. E-commerce will displace conventional methods of business transactions. E-based design and engineering will continue to migrate to low-cost areas of the world.

14.3.2 Relationships

Another abiding theme in the post-war reports is that of roles and relationships of the professions. Throughout the post-war years the professional straitjacket has become tighter, with each profession jealously guarding its autonomy. While interdisciplinary working has been encouraged from Emmerson to Egan, it has yet to be realised. However, the cultural changes to the industry have shaken things to such an extent that the professions now have to consider providing clients with a new framework for delivery services in the future. The next generation of constructional professionals could be provided with a broadly based education and not isolated within what Gann & Salter[13] call professional 'silos'. The Fairclough report[4] on construction research also identified the need to create a career path that provides a broad qualification across the process of design, environmental planning, project planning, construction and beyond, to create a cadre of potential leaders for the industry.

Thus, the most recent reports documented in this volume have created an industrial ecology where ideas of integration can flourish. This concept of integration can find expression in roles and relationships of the professions and how they will influence the construction process and technical practice.

14.3.3 *Performance*

All of the post-war reports sought to encourage improvements in the performance of the industry. This urging continued through to the Egan report with its seven performance improvements demanded of the industry. Will this be the trend for the future? One of the most ubiquitous themes has been that of prefabrication and standardisation. This urge for mass production seems to have natural limits for the future, and the continuance of the rationalist paradigm for mass production has a modernist feel in what has become a post-modern society. Centralised mass production of standard units may offer economies of scale but runs counter to the need to limit the movement of goods around the country or indeed the world. Secondly, other industries, especially ones where customisation is seen as fashionable (fashion, cars, etc.) have pioneered ways of moving from mass production to mass customisation. The construction industry may be expected to follow suit, with the drive to standardise being accompanied with bespoke façades providing individuality to the built landscape. While these wider social and aesthetic environments have not been discussed in terms of performance improvements, they must surely be accommodated in future performance measures which will go beyond an instrumental and narrow range of key performance indicators designed disproportionately to benefit clients.

The dominant paradigm driving the performance critiques of all the post-war reports is one of a rationalist model. Better performance is achieved by rationalising design and construction methods and integrating professional roles with an expectation that these will improve time, cost and quality performance. In the future these issues may be widened to include measurement of performance in terms of quality of life and 'well-being' issues. Certainly, zero tolerance of accidents and deleterious occupational diseases (e.g. stress, overwork and aggression) will be a critical dimension of future performance measures. This will have the effect of eliminating 'dirty work' at the building site; heavy lifting, spraying, exposure to sun, creating dust by sweeping, etc., may not be acceptable, and so most building work will be done in factories and assembled in much the same way as large aircraft. Customisation will take place by a stronger emphasis on façade engineering.

Finally, the features of performance improvement in the past have been decidedly driven by concepts of wealth creation. The future could be driven by the creation of a better sense of well-being; less stress, more leisure, more harmonious professional relationships and, above all, a greater sense of fun and playfulness in our working lives. This is not to celebrate the self-satisfied, somewhat smug, uberclass depicted in Galbraith's The *Culture of Contentment*[14] as by Epicurean values of enjoyment of life without excess. These are real performance issues; the present model is merely what Green[15] calls 'management by stress'. It will perhaps take further cultural and political shifts to move the industry from its present state to one that gives a greater sense of reward, in its widest sense, to those already engaged in the industry and provides a vision that is attractive to idealistic youngsters. It is, however, pertinent to end this text with a positive outlook. A fitting quote is used below which seems to reflect both the historical nature of the book and the optimism shown by the recent Latham and Egan reports. However, if we are to reflect on the 2001 National Audit Office report findings regarding the dissemination of best practice[16], the challenge becomes clear. The *Construction Best Practice Programme* estimates that it has reached only 9% of the working population in the industry through its programme of seminars and workshops. While this is a promising start to the cultural changes demanded of the industry, one is reminded of Churchill's epithet[17]:

'This is not the end. It is not even the beginning of the end. But it is, perhaps, the end of the beginning.'

14.4   References

1   Lloyd-George, D. (1918). What is our task? Speech, 24 November.
2   Flanagan, R., Ingram, I. & Marsh, L. (1998) *A Bridge to the Future: Profitable Construction for Tomorrow's Industry and its Customers*. Reading Construction Forum, Thomas Telford Books, UK.
3   Darbyshire, Sir Andrew (1993) *Crossing Boundaries*. Report on the state of commonality in education and training for the construction professions. Construction Industry Council, London.
4   Fairclough, J. (2002) *Rethinking Construction Innovation and Research*. Department of Trade and Industry, HMSO, London.
5   Gaskell, M. (1989) *Harry Neal Ltd: A Family of Builders*. Granta Editions, Cambridge.
6   Hindle, B. (2001) Structural impediments to construction supply chain development. CIB World Building Congress: *Performance in Products and Practice*, 2–6 April, Wellington, New Zealand.
7   DETR (2000) *Building a Better Quality of Life: A Strategy for More Sustainable Construction*. Department of the Environment, Transport and the Regions, London.
8   Powell, C. (1996) *The British Building Industry Since 1800 : An Economic History*. E&F Spon, London.
9   Hillebrandt, P. (1988) *Analysis of the British Construction Industry*. Macmillan Press, London.
10   Aslop, W. (2001) No beauty in industry. *Contract Journal*, 19 December, 5.
11   Sims, A. (1999) Second opinion. *Building*, 26 March, 3.
12   Rogan, A. (1999) Inter-organisational relations in the construction process. PhD thesis, University of the West of England, Bristol.

13   Gann, D. & Salter, A. (1999) *Interdisciplinary Skills for the Built Environment Professionals*. Arup Foundation. London.
14   Galbraith, J. (1993). *The Culture of Contentment*. Penguin, London.
15   Green, S. (1999) The missing arguments of lean construction. *Journal of Construction Management and Economics*, 17(2), 133–7.
16   National Audit Office (2001) *Modernising Construction*. Report by the Comptroller and Auditor General, London.
17   Churchill, W. (1942) Speech given at Lord Mayor's Luncheon, Mansion House, London, 10 November. Quote taken from *www.winstonchurchill.org*

Index